MAKING NATURAL KNOWLEDGE

Constructivism and the History of Scien⸱

In *Making Natural Knowledge: Constructivism and the History of Science*, Jan Golinski reviews recent writing on the history of science and shows how it has been dramatically reshaped by a new understanding of science itself. In the last few years, scientific knowledge has come to be seen as a product of human culture, an approach that has challenged the tradition of the history of science as a story of steady and autonomous progress. New topics have emerged in historical research, including: the identity of the scientist, the importance of the laboratory, the roles of language and instruments, and the connections with other realms of culture and society. Golinski has written a sympathetic but critical survey of this exciting field of research, at a level that can be appreciated by students or anyone else who wants an introduction to contemporary thinking about the development of the sciences.

Jan Golinski is Associate Professor of History and Humanities at the University of New Hampshire, where he teaches the history of European science since the Renaissance. He has also held visiting appointments at Churchill College, Cambridge University, and Princeton University. He is the author of *Science as Public Culture: Chemistry and Enlightenment in Britain, 1760–1820* (Cambridge University Press, 1992) and of many articles on the history of science. His current work includes editing (with William Clark and Simon Schaffer) a collection of essays on *The Sciences in Enlightened Europe* and investigating the cultural history of weather in the eighteenth century.

CAMBRIDGE HISTORY OF SCIENCE

Editors

GEORGE BASALLA
University of Delaware

OWEN HANNAWAY
Johns Hopkins University

Man and Nature in the Renaissance
ALLEN G. DEBUS

The Construction of Modern Science: Mechanisms and Mechanics
RICHARD S. WESTFALL

Science and the Enlightenment
THOMAS L. HANKINS

Biology in the Nineteenth Century: Problems of Form, Function, and Transformation
WILLIAM COLEMAN

Energy, Force, and Matter: The Conceptual Development of Nineteenth-Century Physics
P. M. HARMAN

Life Science in the Twentieth Century
GARLAND E. ALLEN

The Evolution of Technology
GEORGE BASALLA

Science and Religion: Some Historical Perspectives
JOHN HEDLEY BROOKE

Science in Russia and the Soviet Union: A Short History
LOREN R. GRAHAM

Science and the Practice of Medicine in the Nineteenth Century
W. F. BYNUM

The Foundations of Modern Science in the Middle Ages
EDWARD GRANT

MAKING NATURAL KNOWLEDGE

Constructivism and the History of Science

JAN GOLINSKI

CAMBRIDGE
UNIVERSITY PRESS

PUBLISHED BY THE PRESS SYNDICATE OF THE UNIVERSITY OF CAMBRIDGE
The Pitt Building, Trumpington Street, Cambridge CB2 1RP, United Kingdom

CAMBRIDGE UNIVERSITY PRESS
The Edinburgh Building, Cambridge CB2 2RU, UK   http://www.cup.cam.ac.uk
40 West 20th Street, New York, NY 10011-4211, USA   http://www.cup.org
10 Stamford Road, Oakleigh, Melbourne 3166, Australia

© Jan Golinski 1998

First published 1998

Printed in the United States of America

Typeset in Palatino 10/12 [RF]

Library of Congress Cataloging in Publication Data
Golinski, Jan.
Making natural knowledge : constructivism and the history of
science / Jan Golinski.
p.   cm. – (Cambridge history of science)
Includes bibliographical references.
ISBN 0-521-44471-3 (alk. paper). – ISBN 0-521-44913-8 (pbk.
alk. paper)
1. Science – History.   2. Science – Historiography.
3. Constructivism (Philosophy)   I. Title.   II. Series.
Q125.G63   1998
507'.22 – dc21
97-24028
CIP

*A catalog record of this book is available from
the British Library.*

ISBN 0 521 44471 3 hardback
ISBN 0 521 44913 8 paperback

Contents

Illustrations

Preface

This is a book of uncertain genre, which seems to call for more than the usual amount of prefatory explanation. What follows is a kind of extended historiographical essay, a review of recent writing about the history of the sciences. It is not, however, a comprehensive survey; rather, it is selective and written from a clearly defined point of view. My aim is to explore the implications of what I have called a "constructivist" view of science for the question of how its history is to be written. By a "constructivist" outlook, I mean that which regards scientific knowledge primarily as a human product, made with locally situated cultural and material resources, rather than as simply the revelation of a pre-given order of nature. This view of science has attained widespread currency in recent years, although expressed in a variety of different idioms with varying degrees of explicitness. For historians, as for others, it brings in its train a series of questions: What does such an outlook imply for the history of the sciences? What issues does it open up for historical investigation? What new sources does it suggest historians might be able to use? What questions are posed to history by the constructivist perspective, and in what ways might historical research illuminate, extend, or challenge it?

In proposing answers to these questions, I shall give an avowedly partial survey of recent historical work, choosing to emphasize those studies that seem to draw strength from – or to develop implications of – "constructivism." I shall argue that identification of this theme provides a way to draw together much (though not by any means all) of what historians of science have been doing in the last few years. To make the general tendency explicit helps us to make sense of what has been done and point out directions in which we might go from here. My orientation toward this program is sympathetic, but not entirely uncritical. I shall point out ways in which some of its foundational claims have been questioned, and original approaches modified, as the work has unfolded. It should nonetheless be clear that I think constructivism is worthy of serious attention from those who are interested in the history of the sciences, and that historical study has contributed, and can contribute further, to its development.

Within the limits of my project, I have tried to be flexible in my choice of historical research that can be shown to be relevant to it. Some of the authors whose work I have mentioned may not share my view of their location in the current historiographical landscape. Other mappings of the field would certainly be possible. The value of the view I offer has to be judged by its utility. I have sketched a review of recent historical research that tries to bring into focus some of the most imaginative work of the last few years and to chart a path ahead. In thinking of those who might profit from reading the pages that follow, I have aimed to make them accessible, say, to senior undergraduates who have already studied a little history of science and want to undertake more advanced work. For graduate students in the subject, I offer a guide to some important recent research and a scheme for making sense of its overall direction. It is not, of course, a substitute for study of the monographs and journal articles themselves, but it will help students locate studies that pursue certain methodological themes, and make use of them. I believe the book can serve the same purpose for academic readers in other disciplines, such as general history, philosophy, sociology, anthropology, literary studies, and cultural studies. Readers with a limited amount of time to devote to the history of science will, I hope, be persuaded that they can learn something valuable in relation to their own concerns. Finally, for practitioners of the history of science itself, I offer an opportunity to pause for reflection, to lift our eyes from our immediate research pre-occupations and think about where our subject is going. Not everyone will agree with my view about this – perhaps nobody will agree about everything – but we can benefit, I think, from some discussion of these general issues.

I have chosen to trace the roots of the constructivist outlook to the philosophical arguments of the 1960s and 1970s, surrounding Thomas Kuhn's work and that of the succeeding "Strong Programme" and the "sociology of scientific knowledge." As I shall explain in the Introduction and in Chapter 1, I see the significance of this work as lying in its break with the project of epistemological validation of scientific knowledge – a break that brought in its train a series of novel techniques for the study of science as an aspect of human culture. I propose, in other words, that the uncoupling of historical and sociological inquiry from issues of truth, or realism, or objectivity opened the way to a remarkably productive period in the understanding of science as a human enterprise. Historians, and the others now involved in the interdisciplinary field of "science studies," continue to have reason to be grateful to those who took that step.

Notwithstanding my view of this, I should make it clear that what follows is not a *defense* of constructivism, in philosophical or sociological terms. I am more concerned to see how the constructivist approach can be put to work than to engage in a debate about it on an abstract level. There are, however, a number of very able defenses available, some of which can also serve as introductions to the constructivist outlook. (See, for example, the works of Barnes 1985a; Bloor 1976/1991; Collins and Pinch 1993; Knorr-Cetina and Mulkay 1983; Latour 1987; Mulkay 1979; Pickering 1992, 1995a; Rouse 1987, 1996; and Woolgar, 1988a.) Rather than reiterating philosophical arguments here, I argue implicitly that the best justification of an approach is to show that it can be used productively to generate new knowledge and to deepen understanding. It is nonetheless the case that, in choosing to give serious attention to constructivism, and in ascribing some importance to it in relation to historical understanding of the sciences, I am distancing myself from the recent conservative denunciations of its "fashionable relativism" (Gross and Levitt 1994; cf. Fuller 1995; Lewontin 1995), and even from the more measured discussions that have perpetuated what I believe is a misreading of it as the parent of a postmodernist challenge to the legitimacy of science (Appleby, Hunt, and Jacob 1994). I do not identify the constructivist outlook with a generalized relativism, if by that is meant a determination that all claims to knowledge *are to be judged equally valid*. As I explain in Chapter 1, I take constructivism to be based, rather, on a degree of methodological relativism, which stipulates that all forms of knowledge should be *understood in the same manner* – which is not the same thing.

Many of the sociological and philosophical books on constructivism make use of examples drawn from the research of historians (most recently, Barnes, Bloor, and Henry 1996). This is the first one to reverse the traffic and to explore the implications of the constructivist perspective for historical studies (though Dominique Pestre's recent article in the journal *Annales* [1995] follows a path very close to mine). Looking at it in this connection helps us to see how the program has been articulated and modified by engagement with the specific problems historians face in their work. Rather than portraying history as simply sociological theory put into practice, this account shows how historians have qualified theoretical schemes to accommodate empirical findings that are always more complex than theorists would wish. In the course of this dialectic, the abstract formulations with which constructivism began have been replaced by a more subtle awareness of the complexities of the sciences as creations of human culture. The reader will observe repeatedly, in the chapters that follow, a trajectory that leads from abstract formulations by sociologists and philosophers to the empirically richer and more nuanced accounts of historians. I offer this narrative thread as a reflection of my own experience of the last two decades' work in empirical study of the sciences. My argument is not an antitheoretical one, however. On the contrary, I insist that the course of recent research cannot be understood without acknowledging the critical and continuing importance of theoretical articulations of constructivism. If I had not been convinced of this, I would not have written the book in the way I have.

In tracing the progress of a dialogue between empirical research and theoretical interpretation, I have ended up writing an unusually long essay in historiography. I have however tried to avoid the tendency, sometimes found in that kind of writing, to lay down the law about what is good and what is bad history. I have also tried to avoid the tone of a manifesto or a call to arms. Neither negative criticism nor a programmatic clarion call seems to be necessary, given the wealth of good historical work of the last couple of decades which I have been able to draw upon for examples of concrete achievements. We can take pride, I believe, in what has been accomplished and look with optimism to the future.

Quite early in my thinking about this book, I decided that I could not write a survey, organized along chronological or geographical lines, of the historical picture that has emerged from constructivist studies. Instead, each chapter is organized around a theme that connects historical research with other varieties of science studies. The Introduction and Chapter 1 trace the origins and development of constructivism and outline some of the debates that have characterized its history. Chapter 2 develops the theme of the social dimension of scientific life, which has been brought under renewed scrutiny by constructivism. I describe work on the formation of the identity of the scientific practitioner in early-modern Europe, and on the creation of new disciplinary structures in the so-called second scientific revolution of the late eighteenth and early nineteenth centuries. In Chapter 3, I consider the issue of the locations in which scientific knowledge is produced. I describe work on laboratories, in which materials, instruments, and human skills are concentrated and put to use, and on the fieldwork sciences, which deploy their resources across much more extensive spaces. Chapter 4 looks at science as a linguistic activity, embodied in a variety of different kinds of discourse, from lectures and grant proposals to research papers and textbooks. I discuss how we can understand scientists' use of language as an activity that involves both persuasion and the making of meaning. In Chapter 5, we shall see that the study of science as a practical activity also involves taking seriously the way material resources are used to create knowledge. Two aspects of laboratory work emerge as crucial in this respect: the manipulation of apparatus and practices of visual representation. Chapter 6 moves beyond the laboratory to consider the means by which scientific knowledge acquires authority in general culture, and the implications this has for our understanding of what "culture" is. In this connection, constructivist studies have emphasized the material means by which verbal and visual representations are transported across time and space. The final Coda tackles the question of the responsibilities of constructivism as a form of historical narrative. I consider what forms of writing constructivist history of science might produce, and what the prospects might be for relations with its potential audiences.

Having organized the book thematically, I have drawn for historical examples upon studies that are concentrated in my own areas of competence. The biases and limitations revealed by my choices will be readily apparent to knowledgeable readers. My own scholarly research has been on the sciences of the "long eighteenth century," particularly in Britain. My knowledge of more modern physical and biological sciences is much less comprehensive. And I am quite unable to address the question of how constructivist perspectives could affect our understanding of natural knowledge in the premodern or non-Western worlds. In other respects, my choices of historical research to discuss have been even more arbitrary. I am well aware that there is much relevant work that I have not been able to incorporate, either because of my ignorance or because I did not see how to do justice to it within the small compass and limited number of themes I had chosen. I apologize to authors who may feel slighted in this respect, and (even more) to those who think their work has been misrepresented by my treatment of it.

My most substantial debt of gratitude is to the many colleagues, in the various fields of science studies, from whose work I have learned. For most, a reference in the Bibliography is a scarcely adequate acknowledgment of my indebtedness. I would just add that I feel privileged to have been a witness of the intellectual excitement that has characterized this field in the last two decades.

This book originated in a happy coincidence of my own ideas and those of John Kim, when he was working at Cambridge University Press. After John's departure, Frank Smith and Alex Holzman at the Press continued the encouragement. Owen Hannaway and George Basalla have been most supportive series editors.

The work was begun during my period as a visiting assistant professor at Princeton University, in the spring of 1992. I owe a particular debt to Norton Wise and the other faculty and students in the History of Science Program for their friendly advice.

Just when I was wondering if I would ever have time to write the book, the Dibner Institute for the History of Science and Technology at MIT came forward with a very welcome offer of a resident fellowship in the spring of 1994. I am most grateful to Jed Buchwald, Evelyn Simha, the institute staff, and the other resident fellows, for making my time at the Dibner so profitable.

At my home institution, the University of New Hampshire, colleagues in many departments have fostered and sustained my interdisciplinary interests. Those in the History Department have helped me anchor them firmly in a commitment to historical scholarship, while also giving me the sense of intellectual freedom to undertake a project like this.

Peter Dear, Simon Schaffer, Jim Secord, Steven Shapin, and Roger Smith deserve special thanks for their careful readings of the manuscript and their very helpful comments.

Many colleagues have aided me by sharing their work, by commenting on mine, or just by illuminating conversation. In addition to those mentioned above, I can recall particularly helpful discussions with Mario Biagioli, Jed Buchwald, Harry Collins, Steve Fuller, Dominique Pestre, Larry Prelli, Anne Secord, Miriam Solomon, Maria Trumpler, and Andrew Warwick. No doubt there were others. Anne Harrington, Everett Mendelsohn, and Sam Schweber invited me to participate in seminars at Harvard that were good-humored and very productive. Earlier versions of parts of the book were also presented to audiences at the Boston Colloquium for Philosophy of Science, and at the Dibner Institute. I thank those who commented or asked questions on either occasion, especially Evelyn Fox Keller. I am also grateful to the following for sharing their work with me in advance of publication: Pnina Abir-Am, Jon Agar, Mario Biagioli, Harry Collins, Michael Dennis, Sophie Forgan, Steve Fuller, Graeme Gooday, Dominique Pestre, John Pickstone, Hans-Jörg Rheinberger, Simon Schaffer, Steven Shapin, and Mary Terrall. None of these people should be held responsible for what I have made of their work or their advice.

Introduction: Challenges to the Classical View of Science

[W]hen we tell our Whiggish stories about how our ancestors gradually crawled up the mountain on whose (possibly false) summit we stand, we need to keep some things constant throughout the story. The forces of nature and the small bits of matter, as conceived by current physical theory, are good choices for this role.

Richard Rorty, *Philosophy and the Mirror of Nature* (1979: 344–345)

But, irresistibly, I cannot help thinking that this idea is the equivalent of those ancient diagrams we laugh at today, which place the Earth at the center of everything, or our galaxy at the middle of the universe, to satisfy our narcissism. Just as in space we situate ourselves at the center, at the navel of things in the universe, so for time, through progress, we never cease to be at the summit, on the cutting edge, at the state-of-the-art of development. It follows that we are always right, for the simple, banal, and naive reason that we are living in the present moment. The curve traced by the idea of progress thus seems to me to sketch or project into time the vanity and fatuousness expressed spatially by that central position. Instead of inhabiting the heart or the middle of the world, we are sojourning at the summit, the height, the best of truth.

Michel Serres with Bruno Latour, *Conversations on Science, Culture, and Time* (1995: 48–49)

People who have not come across it before frequently express surprise that there is a subject called "history of science." "Is that a kind of history or a kind of science?" they sometimes ask. "Do historians of science work in a library or a laboratory?" And, "Why would anyone want to study out-of-date science, anyway?"

To those with some knowledge of the subject, these questions no doubt sound naive. But, as I have encountered them repeatedly, I have come to feel that they reflect serious conceptual difficulties surrounding the yoking together of the words "history" and "science." It is not just that the two subjects are usually well separated in educational institutions, but that they seem to be rooted in fundamentally opposed points of view. History is oriented to the past, while science seems oriented to the future; history is connected with humanity, science (largely) with the

1

nonhuman world; history is associated with culture, science with nature; history is thought of as subjective, science as objective; history uses common language, while science uses technical vocabulary; and so on. Because of these common assumptions, the conjunction of "history" and "science" can seem bizarre and confusing.

Even those of us who are now familiar with the history of science should perhaps remind ourselves, from time to time, what a strange hybrid the discipline is. It is worth asking periodically what history of science *is*, what it *can be* as a discipline. This book is about how ideas of the subject have been changing in the last few decades. It argues that the hybridization of history and science has been a remarkably fertile union, giving birth to exciting new ideas about what science is, its role in our culture and society, and the kind of history that is appropriate for understanding it.

In part, these new ideas and approaches have come about through a fundamental reconsideration of the ways of dealing with the past that are embedded in the practices of science itself. Although we sometimes think of science as aimed only toward the future, scientists do have to engage in interpretation and assimilation of the past as part of their work. Practising scientists are continually appropriating the work of their predecessors and orienting themselves in relation to it. They periodically celebrate the work of founders and pioneers of the various scientific disciplines (Abir-Am 1992). It is because their interests in the past do not coincide with those of historians, however, that problems arise. The history of science has had a long struggle to free itself from science's own view of its past.

This is particularly so because of the context in which the subject of history of science originated. When it began, during the eighteenth-century Enlightenment, it was practised by scientists (or "natural philosophers") with an interest in validating and defending their enterprise. They wrote histories in which the discoveries of their own day were presented as the culmination of a long process of advancing knowledge and civilization. This kind of account tied the epistemological credentials of science to a particular vision of history: one that saw it as steady upward progress. The science of the day was exhibited as the outcome of the progressive accumulation of human knowledge, which was an integral part of moral and cultural development. The origins of the history of science lie in this Enlightenment project to advance the standing of natural knowledge by claiming for it a particular kind of history. It is worth considering this legacy briefly, before we discuss the current prospects for the discipline.

For its pioneers, like the eighteenth-century English preacher and chemist Joseph Priestley, the history of science was part of an all-encompassing vision of progress. Human knowledge could be seen to have advanced in a single positive direction, even though its forward motion might sometimes have been delayed (McEvoy 1979). Knowledge of the natural world increased in step with the enhancement of human life in all of its material and cultural aspects – the process that Enlightenment intellectuals called "refinement" or "improvement." Priestley wrote, of his *History of Electricity*, that the "idea of a continual rise and improvement is conspicuous in the whole study"; so that the history of science, thus narrated, "cannot but animate us in our attempts to advance still further" (1767: ii–v). A "philosophical" history of this kind would be more instructive than human or "civil" history, with its disorder and immoral behavior. The history of scientific progress offered readers a "sublime" experience, one provoked by ideas that, "relate to great objects, suppose extensive views of things, require a great effort of mind to conceive them, and produce great effects" (Priestley 1777: 154).

This uplifting vision of the progress of science was integrated with a particular model of epistemology. Priestley, like many of his contemporaries, was an empiricist; he believed that knowledge comprised an association of ideas that derived from the impact of external reality upon the senses. Knowledge was stored up in the mind like marks on the proverbial blank slate. Because ideas, which represented impressions of the external world, could be translated in turn into speech and writing, the stock of human knowledge was constantly augmented.

Even when strict empiricism was brought into question, at the end of the eighteenth century, the epistemological model of the mind as a "mirror of nature" was largely retained, and the history of science continued to be narrated as a story of progress. In the 1830s the man who invented the word "scientist," William Whewell, again argued that the historical development of the sciences followed the path by which the human mind gradually gained representational mastery of external reality. Although Whewell complicated Priestley's version of empiricism, by ascribing an essential role to mental activity in anticipating and structuring experience, the basic notion of knowledge as a representation of the object in the mind of the subject was retained (Brooke 1987, Cantor 1991a). This informed Whewell's view of the historian's task and the kind of narrative he should produce. He wrote:

> [T]he existence of clear Ideas applied to distinct Facts will be discernible in the History of Science, whenever any marked advance takes place. And, in tracing the progress of the various provinces of knowledge which come under our survey, it will be important for us to see that, at all such epochs, such a combination has occurred. . . .
>
> In our history, it is the *progress* of knowledge only which we have to

attend to. This is the main action of our drama; and all the events which do not bear upon this, though they may relate to the cultivation and the cultivators of philosophy, are not a necessary part of our theme. (Whewell 1837/1984: 7–9)

In the second half of the twentieth century, both Whewell's story of progress and its undergirding philosophical assumptions have been subjected to damaging criticism. Historical narratives in which science appears to advance steadily in the direction of greater accumulations of factual knowledge are now widely scorned as "whig history." Priestley's and Whewell's chronicles of the steady progress of discoveries have been revealed as nostalgic retrospectives, like the stories the Whig political historians used to tell about the steady growth of English liberty. Today's historians are more likely to set themselves the goal of understanding the past "in its own terms" (whatever that might mean) rather than in the light of subsequent developments. This has yielded histories in which knowledge, rather than continuously increasing, has undergone radical discontinuities and transformations, and in which what subsequently come to be seen as forward movements are deeply rooted in contexts that are quite foreign from a modern perspective.

Recent criticism has also removed a central philosophical support from Whewell's vision of history – the idea of a universal scientific method. Whewell wrote explicitly to demonstrate the pervasive importance of the method of induction, whereby scientific knowledge was supposedly built up, by generalization from collections of particular observations and experiments, to universal laws. The narrative of progress was designed to display the working through of the inductive method and to recommend its continuing use in science. "It will be universally expected," Whewell wrote, "that a History of Inductive Science should . . . afford us some indication of the most promising mode of directing our future efforts to add to its extent and completeness" (1837/1984: 4). Since Whewell's day, however, many alternative accounts of scientific method have come to be entertained in place of his inductivism. More fundamentally, persuasive arguments have been proposed against the belief that scientists consistently adhere to any single, specifiable method in their research. All the methods proposed have been subjected to stringent critiques, while some philosophers have undermined the whole project of methodology by arguing that human action cannot be understood as a process of following general rules. Meanwhile, sociologists of contemporary science have shown that practising scientists do not appear to be bound by any of the rules of method that have been suggested. No single method that has been articulated seems able to capture more than a part of what they actually do (Mulkay 1979: 49–59). To expect such a method to provide a key to historical development has therefore come to seem quite naive. One account has gone so far as to conclude that the only methodological rule that can be consistently applied to the great scientific innovators of the past is that "anything goes" (Feyerabend 1975).

The effect of these challenges has been to undermine the historical and philosophical assumptions upon which the history of science was originally established. Stories of the long-term incremental progress of accumulating knowledge, under the aegis of the scientific method, no longer command general acceptance. Uprooted from its original philosophical foundations, the subject has nonetheless flourished, aided by a multitude of new intellectual resources and alliances. As the link between whiggish history and classical empiricist epistemology has been broken, new links with other versions of philosophy and history, and with the humanities and social sciences, have been forged. In the process, the history of science has ceased to seem fundamentally different from other fields of human history, although it continues to benefit from a wealth of interdisciplinary connections that other kinds of history sometimes lack. Practitioners of the subject have been able to draw upon the contributions of sociology, anthropology, social history, philosophy, literary criticism, cultural studies, and other disciplines.

It would be impossible to survey all of these contributions here. Instead, in Chapter 1, I shall trace a particular lineage that connects recent historical work back to crucial arguments in the philosophy and sociology of science that surfaced in the 1960s and 1970s, though their roots go back somewhat earlier. I begin with Thomas S. Kuhn, whose *Structure of Scientific Revolutions* (1962/1970) launched a fundamental reexamination of the nature of science. Kuhn's book, as we shall see, was given a forceful if contentious interpretation by David Bloor and Barry Barnes, at the University of Edinburgh, who articulated what they called the "Strong Programme" in the sociology of science in the 1970s. This program, with its founding proposition that science should be studied like other aspects of human culture, without regard to its supposed truth or falsity, was controversial among philosophers and many historians. It nonetheless provided an important inspiration for the field that became known as the sociology of scientific knowledge (or "SSK"), which accrued some impressive empirical case studies and began to influence the work of several leading historians by the mid-1980s. As we shall also see in Chapter 1, SSK faced a significant challenge in the late 1980s from an alternative sociological approach, advocated by the "actor-network" school of Bruno Latour and Michel Callon. The arguments over these different approaches fragmented the community of sociologists of science but, paradoxically, confirmed their influence among historians, and increasingly also among philosophers. By the late 1980s, the constellation of "science studies" disciplines was heterogeneous and riven with arguments, but it was no longer possible to evade the conclusion that the traditional understanding of science had been radically undermined.

History of science took up its position as a participant in this lively and burgeoning interdisciplinary field.

The aim of this book is to map these changes, to survey the areas of research in the history of science that have been transformed by them, and to point to areas that might profitably be developed in future. I use the label "constructivism" to sum up the outlook shared by the sociologists of science whom I discuss and the historians who have been influenced by them. The term draws attention to the central notion that scientific knowledge is a human creation, made with available material and cultural resources, rather than simply the revelation of a natural order that is pre-given and independent of human action. It should *not* be taken to imply the claim that science can be entirely reduced to the social or linguistic level, still less that it is a kind of collective delusion with no relation to material reality. "Constructivism," as I shall characterize it, is more like a methodological orientation than a set of philosophical principles; it directs attention systematically to the role of human beings, as social actors, in the making of scientific knowledge. My argument will be that it has already proven to be a productive orientation for historians, one that opens up many intriguing new issues for historical investigation. The label may be a problematic one – it does have quite different connotations in fields like mathematics and architecture, for example – but it serves my purposes better than alternatives like "the Strong Programme," "social construction," or "the sociology of scientific knowledge," in part because it is not the shibboleth of any particular school.

In tracing the development of the constructivist outlook from the work of sociologists of science, I am aware that I am giving only one of a number of possible lineages. Quite different accounts could very plausibly be proposed. One might, for example, consider how various strands in twentieth-century European philosophy have called into question the model, assumed by Whewell, of the mind as mirror of nature (Rorty 1979). Modern philosophical movements, such as phenomenology, hermeneutics, and poststructuralism, have complicated the subject–object relationship that lay at the heart of classical epistemology. For example, phenomenologists have drawn attention to the ways in which human knowledge is a product of our use of our bodies. Human subjects should not be regarded as detached minds passively contemplating the material world; they learn about it through embodied interaction. Martin Heidegger's philosophy similarly presents all knowledge as the outcome of use – as tools or instruments – of the objects we find around us. It is only by encountering things in the world and using them for our own purposes that we come to know them, in Heidegger's view (Rouse 1987). Meanwhile, hermeneutics and poststructuralism have directed attention toward language, which, they suggest, should not be seen simply as a transparent vehicle for communicating thought. Language is to be grasped in its rhetorical and semiotic dimensions, which go well beyond its function of conveying a message. We use words as much to persuade as to communicate, and, as we do so, much of the meaning of our words escapes from our control. Finally, the social collectivity, ignored in the classical model of subject and object, has come to be regarded as critical for the production of knowledge. One source of this is the later philosophy of Ludwig Wittgenstein, with its claim that language finds meaning by virtue of its use in specific "forms of life" (Bloor 1983). The language we use to describe the world seems to be continuous with our practical activities, an integral part of the actions by which we collectively accomplish our goals.

These philosophical perspectives have yielded important resources for constructivism, and I shall refer to them at various points in what follows. I choose, however, to emphasize the foundations of the movement in studies that were as much sociological as philosophical, and in arguments couched in an idiom that was quite different from that of Continental philosophy. This is because it seems to me that the important move was not so much a change in philosophical outlook as a break of the link that had tethered empirical studies of science to the concerns of classical epistemology. It was primarily by setting aside the attempt to evaluate the credentials of scientific knowledge as rational, methodologically sound, or true to reality that a new space was opened up for research into its creation.

Constructivism was inaugurated by a determination to explain the formation of natural knowledge without engaging in assessment of its truth or validity. This is the position that Bloor called "naturalism": It accepts as "science" for the purpose of study what has passed for such in the context under discussion (1976/1991: 5). It is best understood as a pragmatic or methodological deployment of "relativism" in the service of a comprehensive study of human knowledge. It is a way of screening out the issues of epistemic validity that hinder the understanding of knowledge in its social dimension. Those who have followed this approach have noted how frequently questions of what methods characterize proper scientific research, or what is "good" and what is "bad" science, are intensely disputed in the settings they have studied. It therefore seems inappropriate for the analyst to take sides in these disputes. The ways in which boundaries are drawn between "science" and "pseudo-science," for example, or the way in which hierarchies are created to place "hard" sciences (such as physics) above "soft" (human or social) sciences, are important topics for research; but, to understand the creation of these relationships, the researcher should maintain a neutral stance toward all the contending claims. This methodological principle is also frequently called the *symmetry postulate*.

The postulate is a crucial one because any enterprise that seeks to understand science as a cultural formation has to embrace a wide range of different kinds of knowledge. If we are to understand how astronomy disentangled itself from astrology in seventeenth-century Europe, for example, or how modern Chinese medicine has assimilated elements of traditional and Western therapeutics, it is necessary to put aside attempts to demarcate what is scientific from what is not. We cannot assume that the historical changes we are concerned with were dictated by our own notions about what defines a "science." An open mind on that issue is an important precaution for the historian.

Since the symmetry postulate is primarily motivated by a desire to set aside issues of epistemology, it is unfortunate that it has regularly been attacked as a species of philosophical relativism. One *can* of course assert relativistic claims in a metaphysical or ontological way, saying, for example, that there is no such thing as "truth," or that all beliefs about nature are equally valid, or that there is no "reality" to the material world. But such statements encounter severe difficulties if defended as absolute claims; they lead, as can readily be shown, to self-contradictions. More pertinently, the constructivist outlook does not depend upon them. A pragmatic espousal of "methodological relativism" does not rest upon commitment to all of the absolutes that critics have identified with philosophical relativism (Bloor 1976/1991: 158). Thus, to say that judgments of epistemic validity should not provide the basis of explanations of why certain beliefs are held is not to say that no beliefs can ever be judged valid. To say that nature or reality should not be invoked as determinative of scientific belief is not to deny the existence of the real world or that it has some role in the production of knowledge (cf. Barnes and Bloor 1982; Barnes 1994; Fine 1996).

Understood as a methodological precaution, the symmetry postulate does not imply that questions cannot be asked about the differences between sciences, or between science and other forms of culture. Historians certainly want to explore what distinguishes the practices and social profiles of different scientific disciplines at different times in their development. But these problems can best be approached as topics for open-minded and nonevaluative inquiry. One should not assume that there is only one way for a subject to be "scientific" and only one path of development it can follow. The naturalistic perspective is a way to remind ourselves not to take the development of science for granted.

More can be said about the philosophical implications of the symmetry postulate, but I do not need to discuss it any further now. I simply want to note that, in terms of the development of constructivism, its importance lay in opening the way to a much wider range of empirical studies of natural knowledge in its many different contexts. Cutting the link to the preoccupations of traditional epistemology seems to have liberated naturalistic studies of the sciences, enabling them to explore topics and settings, and to develop approaches, that would not have been favored before. From the historian's perspective, the result has been a thoroughgoing *historicization* of science and all of its associated categories: discovery, evidence, argument, experiment, expert, laboratory, instrument, image, replication, law. These topics have increasingly been approached through research that treats them as historically problematic constructs, in need of contextual explanation, rather than through a priori philosophical analysis (Pestre 1995). As a result, philosophical discussion itself has been obliged to confront the findings of empirical research by sociologists and historians, and has been significantly enriched thereby. It is in this sense we can say, using the formulation that Bloor lifted from Wittgenstein, that interdisciplinary science studies are among "the heirs of the subject that used to be called philosophy" (Bloor 1983: chap. 9).

The transformed relationship between empirical studies of science and philosophical analysis has also been manifested in an increased emphasis on science as *practice*. The pioneers of the Strong Programme tended to take over from previous historical and philosophical studies the assumption that science was best regarded as a body of ideas. Philosophers such as W. V. Quine and Mary Hesse had proposed models of scientific knowledge as a network of interlinked concepts and beliefs. This seemed to explain how it could change in a way that would respond to new experimental findings, but would allow some freedom of choice in how they were accommodated. As constructivist inquiry has unfolded, however, analysts have increasingly tended toward an alternative view of science as a cluster of practices. Rather than a purely intellectual accomplishment, science is viewed as a set of activities in which people engage (Pickering, ed. 1992: 1–26). This shift in perspective owes something to the influence of phenomenology and hermeneutics, but it is more directly connected with the kind of approaches to the social sciences that those philosophies have inspired. Interpretive sociology and anthropology can more readily study what people can be observed to do, rather than what they think. Studies of science that have followed these approaches have tended to abandon the attempt to reconstruct conceptual structures, focusing instead on the practical activities that offer themselves to observation.

Of course, practitioners of the sciences do think, and some constructivist studies have taken a significant degree of interest in cognitive processes (e.g., Gooding 1990). But thinking itself can be regarded as a practical activity, intimately bound up with other kinds of doing; hence, the studies of topics such as the manipulation of materials and apparatus, the production and circulation of images, modeling and analogical reasoning, and all the many ways scientists communicate with one another – conversing, making presentations of results, writing grant applications, composing papers for specialist journals, and so on. Following the drift of naturalism, those exploring the construction of scientific knowledge have generally avoided presuppositions about which of these practices are really crucial and which are of secondary importance. Language use, for example, is not seen as merely communicating what is already known, but as one of the practices by which knowledge is constructed. Thus, the language used at the laboratory bench, in the lecture theater, in the journal article, and in the television documentary are all worthy of study; each serves as the means by which knowledge is produced in that location.

Much more contested than the naturalistic outlook or the tendency to focus on practice has been the issue of how to define the social dimension of scientific knowledge. Although the adjective "social" is frequently used to qualify constructivism, there is in fact no unanimity as to how the social element in the making of scientific knowledge should be specified or what explanatory role should be ascribed to it. The Strong Programme used the symmetry postulate as a wedge to make space for social explanations of the adoption of scientific beliefs. Bloor claimed that these explanations should show how beliefs could be traced to social causes, such as the structure of a scientific community or the interests of those involved. The historical case studies published by members of the Edinburgh school generally made claims that were both causal and *macrosocial*, tying individuals' beliefs to factors such as social class or location in a disciplinary community (Barnes and Shapin 1979).

On the other hand, the sociological studies of controversies, among which those of Harry Collins (1985) were preeminent, made no reference to such large-scale social forces. Collins reiterated the Edinburgh line that neither empirical evidence nor the canons of scientific method could make adoption of particular beliefs logically compelling, so that social causes must be decisive. But the social causes he invoked were on a much smaller scale than the Strong Programme had favored. Collins directed attention to the contingent judgments and negotiations made among small groups of scientific specialists. Controversies were said to be decided by the "core set," a small group of specialists who were closely concerned with the issue in question. Larger-scale social structures and interests (those of class, for example) were not shown to operate in these cases.

As the sociology of scientific knowledge built up further case studies, the trend away from macrosocial explanations was confirmed. Bruno Latour and Steve Woolgar's *Laboratory Life: The Social Construction of Scientific Facts* (1979/1986) was a pioneer in the field of ethnographical studies of specific research laboratories. Karin Knorr-Cetina (1981) and Michael Lynch (1984) were also among the first sociologists to engage in direct observational studies of work practices among experimental scientists. These studies were characteristically focused on a single laboratory; they generally displayed no interest in social forces beyond the laboratory walls. The claim was that interactions among small groups of researchers were no less "social" than large-scale forces, such as classes or political movements. However plausible this is, it certainly suggested a more restricted specification of the social context relevant to understanding scientific practice than had previously been claimed. Along with this went a disregard for the aim of causal explanation. Some of the laboratory studies had been influenced by the outlook of ethnomethodology, which, as an approach to the sociological study of everyday activities, had specifically disavowed causal explanations of social action (Lynch 1993: 90–102, 113–116).

The most radical challenge to assumptions about the social dimension of science came with the articulation of the actor-network approach by Latour and Callon. Latour, in particular, argued that the approach challenges traditional ideas of "the social" no less than it does traditional notions of science. He described how the practices in which scientists and technologists are engaged reconfigure the social world at the same time that they create natural knowledge. It was thus said to be inappropriate to invoke sociological categories to explain scientific practice. Rather than attempting overambitious causal explanations, the analyst should simply "follow" practitioners of science and technology, as they manipulate material, social, and linguistic entities (Latour 1988b, 1992). Significantly, the word "social" was dropped from the subtitle of the second edition of *Laboratory Life*.

These arguments might seem rather abstract when viewed by historians, but that does not mean they have no relevance to the problems of historical practice. On the contrary, historians themselves have grappled with the problem of specifying the various contexts in relation to which science can be understood. They have described the many different social collectivities in which scientists' work implicates them: disciplines, institutions, peer groups, research teams, even national and international communities. Some analyses deal with only one of these levels of the social, but most attempt to integrate two or more of them. Even historical studies that are primarily focused on single institutions or research teams are likely to make some reference to the wider setting. Analysis of larger-scale social forces is usually thought necessary to explain why local practices took the form they did. This is certainly the case with the well-known study by Steven Shapin and Simon Schaffer (1985), which I shall discuss in Chapter 1. They drew upon the reestablishment of stable hierarchical social relations in England after the mid-seventeenth-century civil wars to explain the practices of experimental science in the Royal

Society of London. What happened in the Society's meeting rooms in Gresham College was said to have reflected the broader social arrangements of Restoration England.

Some historians might feel that the issue of the relevant social context in any particular case is an empirical one, to be decided on the basis of the evidence. But there are surely general analytical issues at stake here concerning where we are to look for evidence and how we are to couch our explanations. The questions of how we are to define "the social" in relation to scientific practice, and what is to count as an adequate contextual explanation of that practice, call out for theoretical consideration as well as for empirical study. In pursuing that discussion, close attention to what the sociologists have been saying can be very valuable, though the issue remains very much an open one. Historians, too, have an interest in deciding "what's social about constructivism?" After a more extensive review of the origins of the constructivist approach, it is to that question that I shall turn.

1

An Outline of Constructivism

The only condition on which there could be a history of nature is that the events of nature are actions on the part of some thinking being or beings, and that by studying these actions we could discover what were the thoughts which they expressed and think these thoughts for ourselves. This is a condition which probably no one will claim is fulfilled. Consequently the processes of nature are not historical processes and our knowledge of nature, though it may resemble history in certain superficial ways, e.g. by being chronological, is not historical knowledge.

R. G. Collingwood, *The Idea of History* (1946/1961: 302)

In a break analogous to the one effected by astronomy and physics when they excluded the metaphysical question of the *why* in favor of the positive (or positivist) inquiry into the *how*, the human sciences substitute for inquiries into the *truth* of beliefs (in the existence of God or of the external world, or in the validity of mathematical or logical principles) a historical examination of the *genesis* of those beliefs.

Pierre Bourdieu, "The Peculiar History of Scientific Reason" (1991: 4)

FROM KUHN TO THE SOCIOLOGY OF SCIENTIFIC KNOWLEDGE

Apparently much against its author's wishes, Thomas Kuhn's *The Structure of Scientific Revolutions* (1962/1970), has come to be seen as the harbinger of the constructivist movement. It therefore seems worth discussing Kuhn at some length to discern what he contributed to the development of this approach to science studies. This will inevitably be a selective reading: Quite different pictures of Kuhn have been given elsewhere, for example, in work on the philosophy of science, where he appears in the company of Karl Popper and Imre Lakatos in discussions of rational criteria for theory choice. The Kuhn I shall describe addresses a rather different set of issues, including practical reasoning in science, the role of authority in pedagogy, the nature of controversies, and the definition of scientific communities. These and other matters were drawn out in commentaries on Kuhn by the proponents of the Strong Programme in the 1970s and early 1980s. David Bloor and Barry Barnes,

13

working at the University of Edinburgh, showed how Kuhn's work could yield valuable resources for building a constructivist sociology of scientific knowledge.

I shall follow Bloor and Barnes in their appropriation of certain aspects of Kuhn, but I also want to open up other, more directly historical, implications of his work. While his philosophical leanings are evident, Kuhn was not a sociologist, and his primary disciplinary affiliation was to history. It is paradoxical, therefore, that his direct influence among historians has been at best limited. Outsiders might think of Kuhn as the preeminent historian of science of recent decades, but few inside the field have followed the lead of his schematic model of historical change. Kuhn earned widespread recognition for producing a work of broad chronological sweep and clear philosophical significance, in line with a tradition that stretches back to Priestley and Whewell. But it is also striking, as Steve Fuller has noted (1992: 272), that he has turned out to be the *last* contributor to that genre of "didactic macrohistory of science." We might ponder what this suggests about the implications of his work for historiography. It appears that some components of Kuhn's approach proved radically subversive of the narrative themes traditionally used in macrohistories of science.

As is well known, Kuhn's work was based on a distinction between "normal" and "revolutionary" science. For each scientific discipline, he suggested, a common pattern of maturation would be found, albeit one that occurred in different disciplines at different times. A confused initial period, when research was lacking in coherent organization or aim, was succeeded by the achievement of a "paradigm," which Kuhn defined as a "universally recognized scientific achievement that for a time provides model problems and solutions to a community of practitioners" (1962/1970: viii). Having achieved its paradigm, a mature discipline would enter the phase of normal science, in which research was directed toward developing the paradigm and applying it to solve appropriate problems. Normal science could, however, break down in a "crisis," when the accumulation of anomalies (problems that resisted solution by the accepted methods) would obstruct attempts to continue applying the paradigm. Then a revolution would occur: A new paradigm would come in to succeed the old one, in a process that involved both a shift in the psychological framework within which scientists operated and a transformation in the social organization of their community. After the revolutionary upheaval, normal science would resume, governed by the new paradigm. As examples of revolutions, Kuhn referred to the historical transformations associated with the names of Copernicus, Newton, Lavoisier, and Einstein.

The varying interpretations of Kuhn's work are, to some degree, consequent upon the ambiguities of his crucial term, "paradigm." One commentator discerned no fewer than twenty-one different meanings of the word in Kuhn's text (Masterman 1970). This may exaggerate things slightly, but there is no doubt that ambiguities exist. It is not clear, however, what is the most productive way to resolve them. In the second edition of his book, Kuhn himself distinguished two senses of "paradigm." First was "the entire constellation of beliefs, values, techniques, and so on shared by members of a given community" (Kuhn called this the "sociological" sense). And, second, "the concrete puzzle solutions which, employed as models or examples, can replace explicit rules as a basis for the solution of the remaining puzzles of normal science" (this sense, Kuhn said, was "philosophically . . . the deeper of the two" [1962/1970: 175]).

Constructivist commentators, however, have interpreted the term in a way that merges elements of both of these definitions. They have accepted the basic form of the second definition but combined it with the sociological application suggested for the first. Barnes and Bloor, and more recently Joseph Rouse (1987: chap. 2), have argued that the notion of a paradigm as a concrete exemplar – a model problem solution – suggests a pragmatic alternative to the traditional philosophical view that science is governed by a logical structure of theory, a worldview or Weltanschauung. If paradigms are seen primarily as models, then science appears as an enterprise of practical reasoning governed by accepted conventions rather than by logical deduction from some theoretical structure. This understanding of paradigm, it has been suggested, is both philosophically deeper and sociologically more productive.

There is plenty of warrant for the pragmatic notion of paradigm in Kuhn's text. When he introduced the idea that scientific education operated primarily through conveying concrete exemplars, he noted that such an exemplar "cannot be fully reduced to logically atomic components which might function in its stead" (1962/1970: 11). Rather than a set of stipulations from which deduction might proceed, a paradigm comprises a pattern or model: "like an accepted judicial decision in the common law, it is an object for further articulation and specification under new or more stringent conditions" (23). Analogical reasoning plays a large part in the application of a paradigm, because "a paradigm developed for one set of phenomena is ambiguous in its application to other closely related ones" (29). Kuhn gave two examples of the extension of a paradigm by analogy: the work to apply the caloric theory of heat to a range of chemical and physical phenomena in the eighteenth and early nineteenth centuries; and the monumental enterprise of rational mechanics in the same period, which was devoted to extending the range of Newton's laws of motion to cover the movements of terrestrial and celestial bodies. In each case, as paradigm models were extended to new phenomena, significant experimental and theoretical work was required to make them apply to the new situation. These applications cannot be said to have been specified in advance in the original paradigm; they were, rather, new and original extensions of it.

This view of the way science proceeds is in line with the position that Bloor and Barnes described as "finitism." Bloor defined finitism as "the thesis that the established meaning of a word does not determine its future applications.... Meaning is created by acts of use" (1983: 25). Barnes added, "Finitism denies that inherent properties or meanings attach to concepts and determine their future correct applications; and consequently it denies that truth and falsity are inherent properties of statements" (Barnes 1982: 30–31). Kuhn himself, drawing on the same philosophical resources as Bloor and Barnes, had made the comparison between the application of a paradigm and the attachment of meanings to words. It was Wittgenstein who had first argued that a word derives its meaning, not from an exhaustive definition that can specify all of its possible references, but by a process of extending its usage by analogy from instance to instance. We do not need to know the essence of a "game," for example, or be able to give a complete definition of the characteristics a game must have, in order to learn to apply the term to a family of activities. As we learn, by trial and error, the range of its conventionally acceptable applications, we learn its meaning. Kuhn noted:

> Something of the same sort may well hold for the various research problems and techniques that arise within a single normal-scientific tradition. What these have in common is not that they satisfy some explicit or even some fully discoverable set of rules and assumptions that gives the tradition its character and its hold upon the scientific mind. Instead, they may relate by resemblance and modeling to one or another part of the scientific corpus which the community in question already recognizes as among its established achievements.... Paradigms may be prior to, more binding, and more complete than any set of rules for research that could be unequivocally abstracted from them. (1962/1970: 45–46)

The picture we have, then, is one in which exemplary achievements yield a family of techniques, which, in the course of the paradigm's extension, prove appropriate for solving certain problems. A paradigm is not specifiable as a list of theoretical propositions or methodological stipulations; it is not developed by logical deduction from premises. Rather, the exemplar is learned as a model problem solution and is applied by analogy to what are judged to be similar phenomena. To the extent that the problems presented by new phenomena are solved, the paradigm continues to be adhered to, expanding and modifying its range as time goes on.

As exemplary problem solutions, paradigms are learned as ways of seeing and doing. Quite a lot of the process of scientific education, in

Kuhn's view, consists of imparting unarticulated skills and interpretive dispositions. The perceptual and motor abilities that apprentice scientists have to learn cannot be fully spelled out as a set of rules. Anyone with school experience of learning dissection techniques, for example, will recall how little was taught by the words in a textbook, and how much one depended on the guidance of the teacher, on tinkering, and on comparing one's findings with those of others. Those who have pursued their scientific education further will be familiar with the sense of accomplishment that comes from mastering a particular experimental skill, and will probably agree that there is no other way to achieve it than learning by doing. Kuhn adopted the phrase "tacit knowledge" from the philosopher and physical chemist Michael Polanyi to characterize a large part of what the scientist learns in the course of being trained in a particular paradigm. Polanyi had argued that scientific practice requires the learning of substantial unspoken skills: "When we accept a certain set of presuppositions and use them as our interpretative framework, we may be said to dwell in them as we do in our own body. . . . [A]s they are themselves our ultimate framework, they are essentially inarticulable" (Polanyi 1958: 60). Education in the sciences appears less like a process of logical programming than an apprenticeship in a traditional craft. Hence, Jerome Ravetz's (1971) vision of science as "craftsman's work" in a book that took Polanyi's perspective as its point of departure.

The appeal of this view for Barnes and Bloor was that it made science look more like other aspects of culture. As a kind of craft, science appears more open to the naturalistic approach to understanding its construction; the epistemological barriers around it are lowered. It is then plausible to regard scientific beliefs as being inculcated by relations of authority within educational institutions and sustained by conventions in specific communities. Echoing the statement of Polanyi (1958: 53) that "to learn by example is to submit to authority," Barnes wrote, "paradigms, the core of the culture of science, are transmitted and sustained just as is culture generally: scientists accept them and become committed to them as a result of training and socialization, and the commitment is maintained by a developed system of social control" (1985b: 89).

Such a view opens the way to a wide range of possible empirical studies of scientific education, of the workings of institutions, and of the ways in which the authority of science is maintained in the wider society. We shall be considering some of these studies, both sociological and historical, in subsequent chapters. In analyzing these topics, constructivist research has broken with what had been the prevailing approach to understanding the social relations of science. Traditionally, science had been seen as maintained by certain institutions that might be necessary for it to flourish but did not affect the content of what was believed. The constructivist outlook suggests, however, that science is shaped by social relations at its very core – in the details of what is accepted as knowledge and how it is pursued. Kuhn can be read as endorsing one position or the other, depending on the reader's orientation. For the proponents of the Strong Programme, the more significant reading was the more radical one, in which science was seen as social through and through. This reading built upon Kuhn's comments to the effect that paradigms are integral to the definition of scientific communities.

Kuhn used the researchers on electricity (known as "electricians") in the mid-eighteenth century as an example of a scientific community defined by common acceptance of a paradigm. Prior to Benjamin Franklin's work in the 1740s, he noted, there were numerous views about the nature of electricity, but most were held by only a single experimenter and derived from a limited subset of all the known experiments. Franklin's research, focused on the Leyden Jar (a device that could store a remarkably large quantity of electric charge), led to his articulation of a single fluid theory of electricity. The idea was that neutral bodies contained a certain proportion of electrical fluid to normal ponderable matter, and that they could be charged positively or negatively by adding or removing fluid. The theory provided reasonable explanations of conduction and neutralization phenomena; it yielded a plausible account of how the Leyden Jar could be charged; and it could even be extended to cover most (though not all) attractions and repulsions between charged bodies. According to Kuhn, the "Franklinian paradigm . . . suggested which experiments would be worth performing and which . . . would not" (1962/ 1970: 18). The theory, which offered an exemplary solution of an outstanding problem, ended the confusing debate between rival interpretations and gave the electricians a coherent plan for further research.

Kuhn's claim that Franklin's theory deserves to be designated a "paradigm" has been questioned by some subsequent historical research. But I am not concerned with that issue as much as with the way Kuhn identified the achievement of a paradigm with consolidation of the social community of researchers. He noted that "the emergence of a paradigm affects the structure of the group that practices the field." Following Franklin's work, "the united group of electricians" achieved a "more rigid definition" (1962/1970: 18–19). Those who were not willing to adopt Franklin's model were condemned to work in isolation. It seems that group solidarity was the result of consensus acceptance of the paradigm, a coherent social group coalescing around recognition of a specific scientific accomplishment. Kuhn did not talk, at this point, about what might be called the "external" sociological dimensions of the group. He did not describe its place in scientific or educational institutions or its members' positions in society. The implication is that these factors were of secondary importance to the social cohesion that flowed from achievement of the paradigm.

In the "Postscript" to the second edition of his book, however, Kuhn seemed to qualify this point. He noted that there was a circularity inherent in his assumptions that, "A paradigm is what members of the scientific community share, *and*, conversely a scientific community consists of men who share a paradigm" (1962/1970: 176). The circularity was a source of problems that could be avoided, he suggested, if the investigation were to begin with "a discussion of the community structure of science." An objective study of scientific institutions could provide a grid upon which the social influence of various paradigms could be mapped. Accordingly, Kuhn referred to research done by sociologists working in the tradition of Robert K. Merton, who had charted some of the institutional features of modern science. They had drawn attention to such factors as uniform educational experiences, a high degree of unanimity of professional judgment, and publications and organizations serving specific disciplines, as the external conditions that sustained the social cohesion of the scientific community.

From the standpoint of the Strong Programme, however, to attempt to identify social structures that were independent of scientists' commitment to particular forms of practice was to betray one of the fundamental insights suggested by the notion of a paradigm. The circularity of defining a social group and a form of practice in terms of one another need not be a vicious one. Rather, it comes close to what Barnes and Bloor took Wittgenstein to have meant by a "language-game" or "form of life." For them, it was precisely Kuhn's willingness to explore the kinds of communities that form around an exemplary model of practice that gave his work its appeal. These groups would be expected to be narrower than scientific disciplines, and certainly narrower than the population of all professional scientists studied by the Mertonian sociologists. Kuhn was suggesting that it is essential to study smaller-scale groups whose identity is bound up with allegiance to a specific mode of practice. While it might be possible to describe the social locations of these groups in institutional or disciplinary terms, their identity was not to be *defined* in such terms. Rather, their defining feature was their consolidation around a particular way of doing science. Only with this kind of analysis could one hope to show how social relations penetrate to the core of scientific practice.

Kuhn himself had noted that defining communities in terms of external institutional criteria could not help with isolating the interdisciplinary groups that come together to study particular phenomena. He gave the example of the "Phage Group" – the consortium of biochemists, microbiologists, and geneticists that pursued the study of bacterial viruses in the 1940s and 1950s (1962/1970: 177). Barnes talked about the research "sub-cultures," which Kuhn's work had illuminated. Kuhn had shown, he claimed:

[J]ust how profound and pervasive is the significance of the sub-culture in science, and the communal activity of the organised groups of practitioners who sustain it. The culture is far more than the setting for scientific research; it is the research itself. It is not just problems, techniques and existing findings which are culturally specific; so, too, are the modes of perceiving and conceptualising reality, the forms of inference and analogy, and the standards and precedents for judgment and evaluation which are actually employed in the course of research. (Barnes 1982: 10)

The place in Kuhn's analysis where this came out most clearly was in his treatment of the controversies that occur in transitions between paradigms. During scientific revolutions, he claimed, the proponents of competing paradigms engage in debates that are unsatisfying and ultimately likely to prove inconclusive. It is rarely possible for upholders of the new paradigm to provide definitive proof of its superiority to those who continue to defend the old one. This is because the paradigms are themselves the sources of outstanding problems, techniques for solving them, and standards for assessing the solutions. The perceptual skills of scientists are refined to pick out the data that are meaningful for their particular paradigm. Furthermore, because commitment to a paradigm involves acquiring skills in the use of certain instruments, it may be said that adherents to different paradigms are living in different perceptual worlds. We should not expect it to be possible for them to agree on a single set of data against which both paradigms can be assessed. Instead, new paradigms yield new data which are simply outside the range that the previous paradigm would have considered meaningful.

Since, Kuhn said, "there are no standards higher than the assent of the relevant community," there are no *neutral* standards, external to the paradigms, against which they can both be measured. This is a situation of "incommensurability." "When paradigms enter, as they must, into a debate about paradigm choice, their role is necessarily circular. Each group uses its own paradigm to argue in that paradigm's defense" (Kuhn 1962/ 1970: 94). In these circumstances, logical proof of the superiority of one or the other paradigm is not to be expected. "To the extent, as significant as it is incomplete, that two scientific schools disagree about what is a problem and what a solution, they will inevitably talk through each other when debating the relative merits of their respective paradigms" (109). Helping himself to Wittgenstein's phraseology, Kuhn remarked that the choice between competing paradigms was one "between incompatible modes of community life" (94).

The Wittgensteinian phrase suggests how many of the values of each group are at stake in a controversy between paradigms. It also implies that the dispute may be understood in terms of the social organization of the subcultures involved. This was one of the most important lessons learnt from Kuhn by constructivist sociologists and historians. The two decades after the appearance of his work saw numerous studies of scientific controversies, which seemed to occur rather more frequently than Kuhn's picture of occasional revolutions had suggested. These studies, of which those conducted by Harry Collins (1985) were – in their own way – "paradigms," confirmed Kuhn's observation that in such disputes fundamental values are exposed. It was found that participants articulate what are normally unspoken rules of method while assessing disputed phenomena. Kuhn's perception that what is normally "tacit knowledge" comes to the surface in controversies proved to be one of the most fertile insights of his work. It also proved possible to expose how "modes of community life" are involved. Participants in disputes frequently deploy social assumptions about the expertise or reliability of other scientists, or about the propriety of their ways of collaborating or communicating. At times, the debates do indeed seem to be about how research communities should be organized, or in general how science should be conducted.

Historians also began to pursue studies of controversies in the 1980s. Martin Rudwick (1985) and James A. Secord (1986) both used the rich documentation of Victorian geology to show how consensus acceptance of theories concerning the ordering of strata had only been achieved as the outcome of prolonged disputes among experts. Steven Shapin and Simon Schaffer's *Leviathan and the Air-Pump* (1985) brought the controversy-studies technique to bear on what they portrayed as a turning point in the origins of modern experimental science. The dispute between Robert Boyle and Thomas Hobbes in the 1660s was focused on the air pump and the experimental phenomena, such as air pressure, elicited with its aid. But, as Shapin and Schaffer demonstrated, by denying credibility to Boyle's claimed "matters of fact," Hobbes was implicitly calling into question the whole "form of life" that made experimental knowledge possible. By taking Hobbes's objections seriously, it was therefore possible to delineate Boyle's techniques of constructing experimental knowledge, from the outside as it were, in a way that did not take the experimental method for granted. Experiment was shown to rely for its success on a culture that comprised certain material instruments, certain rhetorical and literary techniques, and a certain form of social organization. The decorum that surrounded experimental practice within the Royal Society (from which Hobbes was excluded) was shown to have paralleled the means adopted to ensure civil peace in Restoration England at large. Shapin's and Schaffer's was thus a sociological study, one anchored in the empirical studies of controversies, rather than a traditional social history: It began with disputes over "technical" facts and argued *outward* to the broader issues that were revealed to be at stake, rather than arguing from social context *inward* to technical content. The authors' debt to the controversy studies, and – through them – to Kuhn's account of the incommensurable debates between competing paradigms, is clear.

To recapitulate, we can distinguish three features of Kuhn's analysis that the constructivist approach has followed up. First is the realization that forms of scientific practice are learnt through relations of authority and maintained by the social discipline that sustains consensus in scientific communities. As M. D. King wrote, in a perceptive discussion of Kuhn, this makes science out to be "a system of traditional authority" (King 1980: 103). Second is the picture of scientific practice as governed by adherence to certain model problem solutions, in which theoretical concepts, methods, and commitments to particular instruments are implicit. These implicit values are not, however, fully specifiable in the form of explicit rules. Hence, application of the models to new situations is not determined by logical deduction but by pragmatic judgment. Scientific research is not like the logical step-by-step reasoning of a computer; it is more like the skilled judgment practised in a traditional craft. And third is the insistence that some of the most important values governing scientific practice are quite local, frequently being specific to subcultures considerably smaller than all of the practitioners of a discipline (all physicists, for example). These local cultural values are tied to forms of social life and can be found articulated to some degree in situations of controversy.

These are the fundamentals of a constructivist interpretation of Kuhn. They appealed to the proponents of the Strong Programme as reinforcing the arguments for a "naturalistic" approach to science studies. As science came to be seen as "a system of traditional authority," a craft activity, and a form of "local knowledge," the plausibility of treating it on the same level as other aspects of culture was enhanced. Naturalism or methodological relativism looked like the appropriate stance to adopt. It seemed necessary to shelve issues of truth or validity in order to understand the forms that scientific knowledge took in different cultural settings.

There were also, however, fundamental concerns of the Strong Programme that did not derive from Kuhn. The most important of these was an interest in the social causation of scientific beliefs. Bloor explained that he and his colleagues wanted to explain how scientific knowledge was *caused* by social conditions. This was, in fact, the first of the four canonical tenets that he specified for the Strong Programme (the other three being "impartiality" and "symmetry" – now usually collapsed into the "symmetry postulate" – and "reflexivity," the requirement that the kind of explanation given by the sociology of knowledge should also apply to itself [Bloor 1976/1991: 7]). He stressed that there was no implication that social circumstances *alone* could explain beliefs: "The strong programme says that the social component is always present and always constitutive of knowledge. It does not say that it is the *only* component, or that it is the component that must necessarily be located as the trigger of any and every change" (166). Nonetheless, the whole drift of the program was toward explaining scientific belief by reference to the social realm, and, for Bloor, it was vital that these explanations should be causal in form. Unless sociological explanations could exhibit causes of particular beliefs, the sociology of knowledge could not lay claim to the scientific status that, Bloor insisted, was crucial to its success. This would mean that science was incapable of knowing itself – "a most striking oddity and irony at the very heart of our culture" (46).

Although Kuhn's concept of a paradigm had a very prominent social dimension, it did not offer anything much by way of fulfilling this aim. Kuhn did not present his work as exhibiting social causes for scientific knowledge; nor did his accounts refer very much to events in the world beyond the scientific community. He made use of the dichotomy between "internal" technical factors and "external" social ones, which was popular in historiographical discussions at the time, to downplay the importance of "external" causes in the revolution initiated by Copernicus: "In a mature science – and astronomy had become that in antiquity – external factors . . . are principally significant in determining the timing of breakdown, the ease with which it can be recognized, and the area in which, because it is given particular attention, the breakdown first occurs. Though immensely important, issues of that sort are out of bounds for this essay" (Kuhn 1962/1970: 69; cf. Shapin 1992). Practitioners of the Strong Programme were therefore obliged to look elsewhere for theoretical resources to pursue this aspect of their project.

Two strategies were explored. The more important one was to identify the "interests" held by individuals or small groups of scientists and to use these to explain the choices and judgments they made. This was, in part, inspired by the work on "knowledge-constitutive interests" by the German philosopher Jürgen Habermas (Barnes 1977), but the focus was specifically directed at the influence of social motives on science. A prominent example was the work of Donald MacKenzie and Barnes on the controversy between "biometricians" and Mendelians in early-twentieth-century British genetics (1979). After considering the extent to which the dispute could be explained by awareness of different sets of evidence, or differences in professional training, on each side, MacKenzie and Barnes concluded that wider social interests were involved. Karl Pearson, who defended the biometricians' theory of smooth and continuous evolutionary change, was said to have seen this as a way of legitimating his eugenic aims for the rational improvement of the human race. It seemed more plausible that deliberate intervention could succeed in modifying human characteristics if those characteristics were subject to continuous variation in nature. William Bateson, who took the Mendelian line that evolutionary change could occur in sudden jumps, opposed eugenics and despised the secular middle-class interests that it served. MacKenzie and Barnes claimed that an explanation of the controversy had to make reference to the conflicting social aims of its leading protagonists.

The invocation of interests in a causal role has had a mixed fate in constructivist studies since the Strong Programme. On the one hand, the aims or goals of actors were often invoked to explain their beliefs, though they were usually not formally identified with specific causal "interests." Andrew Pickering, for example, in his *Constructing Quarks* (1984), used this kind of explanation in connection with episodes from the recent history of high-energy physics. He proposed that the physicists he discussed were guided by an attitude of "opportunism in context"; they made the decisions they did because they sought to employ their particular specialist skills through developing new areas of work (Pickering 1984: 10–13, 187–195, 403–414). This seems to be an explanation along the lines suggested by Shapin, who argued that scientists might be shown to be guided by complexes of technical skills and competencies that "represent a set of vested interests *within* the scientific community" (1982: 164–169). There is no stipulation that the relevant interests must *always* be of an "external" kind; they might more frequently be expected to be specific to particular disciplines or subcultures.

On the other hand, however, some sociologists have subjected the category of interests to stringent critique, arguing that they cannot be identified as stable entities of the kind that explanations could be based upon. Steve Woolgar has argued that individuals are constantly respecifying their interests, which serve a basically rhetorical function of justifying particular actions (1981). Others have proposed that causal explanations of any sort should not be the aim of the social analysis of human action (Lynch 1993: 57–60, 65–66). Pickering himself has recently written of aims and interests as liable to transformation in the course of engagement in scientific practice. Individuals may modify their goals and interests as they grapple with the resistances that the material world sets up in opposition to what they try to accomplish (Pickering 1995a: 63–67, 208–212). These arguments do not appear to have been widely attended to by historians, but many historians have participated in a general movement in the human sciences in recent years that has led away from attempts to provide causal explanations of events and toward interpretive accounts of human action. For this reason, although informal identification of the goals and aims of historical subjects remains part of historians' practice, there is less willingness to specify interests as independent causes of actions.

The other strategy of causal explanation explored by the Edinburgh school employed categories of social cohesion and differentiation. Bloor referred to this kind of explanation in discussing Kuhn's account of scientific revolutions. Kuhn had sketched the stages of a revolution, with accumulating anomalies in the old paradigm leading to a sudden "gestalt switch" to a new one; but this was scarcely mentioned by Bloor or Barnes. As Bloor saw it, slightly vague analogies to political revolutions and the use of psychological notions were no substitute for a thoroughgoing sociological study of the shift from one paradigm to another (1983: 142–143). Such a study, he suggested, would dispel the idea that a simple accumulation of anomalies could of itself produce a crisis. Far more likely would be a situation where not everyone in the paradigm community would even recognize the purported anomalies, or agree that they were building up in a way that threatened the paradigm. Given the flexibility with which paradigms can be adapted to new phenomena, some members of the community could always say that what others took to be an anomaly just required a new modification of the paradigm. Given sufficient creativity and resourcefulness on behalf of its defenders, the existing paradigm could be maintained indefinitely. Other responses, such as ignoring purported anomalies completely and pressing on with development of the paradigm in other areas, would also be possible.

Why, then, should a paradigm ever change? Bloor insisted that the answer would require attention to the social characteristics of the paradigm community and the balance of forces within it. Anomalies themselves would be powerless to change anything unless they had champions who insisted on their significance and proposed a new paradigm to explain them. What happened then – whether a revolution occurred or the anomalies were simply ignored – would depend on the social configuration of the community. Bloor developed a typology of communities, derived from the work of the anthropologist Mary Douglas, to try to predict the outcome of conflicts between defenders and critics of a dominant paradigm. He suggested that measures of the stratification and relative openness of a community could be used to predict how receptive it would be to anomalies and how willing to replace a dominant paradigm to take account of them (Bloor 1978, 1983: 138–149).

Bloor's "grid-group" typology of paradigm communities has not been widely adopted, perhaps because it seems too schematic to historians who may not share his goal of causal explanation of scientific change. Nonetheless, his reading of Kuhn indicates a direction in which others have found it useful to go. Attention is shifted from the anomalies as such to their construal by actors with particular purposes. Rather than asking, "What were the anomalies?" it seems more appropriate to ask, "Who was claiming there were anomalies, and why were they successful in getting others to agree?" This form of question opens the door to inquiry into the distribution of resources within the scientific community. As with his picture of normal science, then, Kuhn's contribution to the understanding of scientific change can be read, not as a "theory" as such, but as a set of pointers toward productive lines of inquiry. Just as he directed attention toward the structures of authority and exemplary models responsible for sustaining scientific disciplines, so he highlights the emergence of anomalies as a factor in scientific change. It is, however, crucial that anomalies not be regarded as capable of acting independently; they can only be tools in the hands of human actors who use them to advance specific ends. This is an important point that emerges from the Strong Programme's reading of Kuhn.

This reading is, as I have noted, a partial one. There seems little doubt that it was one with which Kuhn himself did not agree. His own work after *The Structure of Scientific Revolutions* lay not at all in the direction that the Strong Programme's appropriation of that book indicated (e.g., Kuhn 1978). It was, nonetheless, with this kind of interpretation of Kuhn that constructivist work in science studies largely began. The themes this work took up included: the local specificity of scientific subcultures; their structuring around particular models of practice and implicit methods; the logically underdetermined, open-ended nature of scientific work; the role of pedagogical authority and social control in sustaining disciplines; and the possibility of change through mobilization of anomalies by groups that attack the dominant paradigm.

As we shall see, these themes have been developed in studies that have generally been narrowly focused, both chronologically and geographically. Since Kuhn, "microhistories" of science have become the norm, in which a single controversy, institution, discipline, or research program, is scrutinized over a limited period. Although partly reflective of trends in other fields of history, such as the increased prevalence of local studies in social and cultural history, this narrowing of focus is also the result of concentration on the themes that constructivism found significant in Kuhn. It was, however, prefigured by Kuhn's own failure to follow through his gestures toward a macronarrative of the history of science. His work derived much of its authority, especially among sociologists and philosophers, from its sweeping command of the history of the natural sciences from ancient times to the twentieth century; and it is clear that Kuhn was proposing a new form within which a large-scale macronarrative could unfold. Readers might have expected to encounter a chronological account of the major revolutions and the successive paradigms in each discipline. But such expectations were frustrated: Kuhn never provided such a narrative. Nor is it possible to extract unambiguous answers if one asks the necessary questions of his text. When introducing the notion of a paradigm, Kuhn mentions the works of such great scientists as Copernicus, Newton, Lavoisier, and Einstein (1962/1970: 6–7); but he later adds numerous other names, such as those of Dalton and Franklin. In his 1969 "Postscript," he noted that he intended the term "revolution" to apply even to frequently occurring changes in communities that might number no more than twenty-five specialists (181). Yet the term also clearly applies to much larger-scale events, in which disciplines as a whole are transformed or created. In view of this variation, it is scarcely surprising that no clear picture emerges of the succession of revolutions and paradigms, even in a single discipline. And no writer since Kuhn has produced any such large-scale picture.

Since Kuhn, in fact, the historical macronarrative has undergone rapid decline. Kuhn's injection of discontinuities into the smooth flow of progress envisioned by Priestley and Whewell undercut the validity of the whiggish story of scientific development. But no one has provided a plausible alternative narrative that could sustain a chronologically extended account. Instead, historians have focused down on the specific settings of scientific practice, many of them exploring with sociologists the constructivist legacy of Kuhn's work. Shapin and Schaffer's controversy study is a brilliant realization of certain Kuhnian themes, but one that lacks a diachronic perspective. The authors describe and analyze a controversy, which is proposed as foundational for modern experimental science, but there is little sense of development or resolution, and certainly no account of how the experimental "form of life" was sustained chronologically. The task of synthesizing the results of this kind of local study into new large-scale historical narratives remains uncompleted.

WHAT'S SOCIAL ABOUT CONSTRUCTIVISM?

In the late 1970s and early 1980s, the development of the sociology of scientific knowledge (SSK) at first proceeded along lines that might have been predicted by the advocates of the Strong Programme. Case studies of contemporary science were presented by researchers such as Pickering (1984) and Trevor Pinch (1986a), which seemed to instantiate many of the claims articulated by Bloor and Barnes in their reading of Kuhn. Scientific practice was shown as open-ended and underdetermined, scientists not being compelled either by logical deduction from existing beliefs or by unambiguous evidence to develop their ideas in a particular direction. Instead, they were found to be making practical judgments that could be related to the local subculture in which their resources and skills were invested and their specific aims pursued. In controversies, the texture of these (normally hidden) commitments was brought to the surface and exposed to view.

New themes were also explored, especially by those who pursued ethnographic studies of the work of particular laboratories, for example, Latour and Woolgar (1979/1986), Knorr-Cetina (1981), and Lynch (1984). Laboratory studies illuminated both the specific local practices of experimentation and the means by which research findings were communicated to the world outside. Important resources for constructivist history of science were developed in the course of this work. There were also, however, significant debates, which were to lead to the fragmentation of SSK by the late 1980s. The crucial question concerned the characterization of the social realm that was used to explain scientific practice and the nature of the explanation being offered. Bruno Latour was the most outspoken advocate for the view that explaining scientific practice, and the extension of its effects through society, required a drastic revision in the categories that the sociologists of scientific knowledge were using. What was required, he proposed, was to recognize the status of non-human entities as social actors. The dispute sparked by Latour's proposal was fierce and wide-ranging; it also has important implications for how the history of science is to be written.

Among the most influential of the pioneering case studies of contemporary science were those of Collins (1985). Much of his work focused on replication, a process that had previously escaped substantial scrutiny, though it was crucial to traditional notions about the universality of scientific knowledge. The standard assumption is that, since science is inherently universal, replication is "in principle" possible always and everywhere. Collins took a critical and empirical approach to this issue. He explored some situations where an experimental finding was agreed to have been replicated and some where replication remained contested. One of his studies concerned attempts made in British laboratories to replicate the construction of a certain type of laser (the "TEA laser"), which was reported as having been built in Canada in 1970 (Collins 1985: chap. 3). Although it was fairly straightforward, in this case, to decide when the goal of a working laser had been achieved, actually getting to that point was far from straightforward for any of the groups involved. Collins showed that a considerable amount of tacit knowledge had to be conveyed in order for the laser to be reproduced in a new site. Investigators visited one another's laboratories to elicit information not included in printed descriptions of the device, and to set eyes on their colleagues' versions. Some skills could only be transferred in the persons of researchers or technicians who had spent time in the laboratories where there were working instruments. Replication, in other words, was shown to require the transfer of a good deal of the subculture that had surrounded the original production. The achievement of a replica was the outcome of skilled judgment and craft rather than of following a set of rules. As Collins put it, there is no "algorithmic recipe" (1985: 143) for successful replication – a conviction that underlies his skepticism about the claims that computers can be programmed to conduct experimental research (Collins 1990).

In other case studies, Collins explored how replication could fail. This was the case with attempts to confirm Joseph Weber's claims to have detected gravity waves in the early 1970s. In considering this case, Collins highlighted what he called "the experimenters' regress" – a consequence of the impossibility of specifying in advance all of the conditions that must be met for an experiment to be conducted successfully (1985: chaps. 4, 5). Experimental results can only be assessed by reference to a complex set of contextual factors; they are accepted only if the methods are deemed proper, the apparatus appropriate, the investigators competent, and so on. Verdicts on all these questions stand or fall together, and an experiment cannot be accorded a decisive outcome independent of such complex, situated judgments. What this implies is that, because any subsequent reenactment will always differ in some respects from the original experiment, there is always room to argue that it differs in some relevant respect, which makes it not a fair comparison with the original. Although the individual who performs the reenactment may claim to have replicated the earlier experiment correctly, either confirming or falsifying its result, the original experimenter can always assert that the second experiment is different in some relevant way and hence not a valid replication. Assessments of whether a result has been replicated are thus always matters of judgment. Sufficient differences between two versions of an experiment can always be found by a critic who wishes to deny that a proper replication has been achieved. In principle, Collins claimed, such a dispute could be continued indefinitely.

In fact, disputes are usually not very prolonged, and many claims to new knowledge are accepted without any debate. This is due, Collins argued, to the existence of social means for averting and resolving controversies. Scientists do not always challenge other experimenters' claims because, most of the time, they repose trust in the competence of their colleagues. The social links that bind scientific subcultures together are essential conditions for the production of consensually accepted knowledge (just as, in Kuhn's analysis, relations of authority are basic to normal science). Most scientists, most of the time, live their lives within a supporting matrix of trust. It is only when the trust breaks down that its social mechanism is exposed to view. Controversies may not typically be very lengthy, or even very frequent, but Collins accords them prime place in the sociology of science because they reveal the relations of authority and credit which are concealed in knowledge that has become widely accepted. In Collins's striking metaphor, they enable us to see "how the ship gets into the bottle."

Situating scientists in networks of trust was the main way in which Collins provided them with a social context. The suggestion was that formal institutions were of less importance than the informal connections among members of the scientific community, for example, the "core set" of researchers that communicated with one another about a specific controversial issue (Collins 1985: 142–145). The social relations Collins invoked were largely internal to scientists' networks; he did not make reference to the larger-scale interests, concerning such factors as class position or social stability, favored by the Edinburgh school. Recently, Shapin has explored the importance of networks of trust in relation to scientific knowledge in the seventeenth century, in his *A Social History of Truth* (1994). Like Collins, he claims that the perpetuation and extension of scientific knowledge is dependent upon practitioners' trust in one another's factual statements. Unless a large portion of what is reported by researchers is accepted on trust by their colleagues, there can be no substantial body of natural knowledge. In the period when the scientific community was in the process of formation, however, these relationships cannot be supposed to have existed ready-made and isolated from society at large. Shapin therefore argues that relations of trust and credit had to be built up among natural philosophers in the seventeenth century, and that this was done by exploiting common assumptions about gentlemanly decorum and credibility. The image of the gentleman as a reliable truth-teller was a valuable asset, in Shapin's view, for the construction of a community of experimental natural philosophers who could trust one another's word. In this way, Shapin suggests that the Edinburgh school's openness to features of the wider social landscape might well remain important to historians, at least while they are concerned with the period in which the modern scientific community was being constructed.

In general, however, the drift of SSK was away from a focus on the social world at large. In some respects, Collins himself exemplified a general narrowing of the scope of analysis toward the local site of scientific activity. The *laboratory* was identified, in many studies, as the crucial setting for the production of natural knowledge and the place where the social dimension of that knowledge might be best discerned. In the laboratory, it was claimed, facts of nature are produced by using the special resources of equipment and skills concentrated there. Experimental knowledge, at its point of origin, is quite private, though none the less social for that.

In a number of studies, this private space of the laboratory was penetrated by the figure of the ethnographer. Like an anthropologist studying an alien culture, the observer of the laboratory took up a carefully deliberated position, balanced between naive ignorance and complete acceptance of the "natives'" way of seeing things. In their influential study, Latour and Woolgar (1979/1986) made much play with the figure of the anthropologist, joking about whether the "natives" in the San Diego biochemical facility they studied were engaged in sacrificing animals to placate some angry god or in examining their entrails for the purposes of augury. Latour subsequently acknowledged that this "very naive version of the naive observer" (1990: 146) was something of a fictional device, but the fiction is one that other analysts have also found helpful. Shapin and Schaffer introduced their study of Boyle, Hobbes, and the experimental way of life by proposing to "play the stranger" (1985: 6) in order to delineate a culture in which the meaning and validity of experimentation was still being forged.

Jokes aside, then, the stance of the ethnographer can be a valuable one for sustaining a focus on the observable externals of practice and the local specificity of the laboratory site. Two basic points have been consistently maintained by laboratory studies framed in these terms. First, scientists at work do not seem to be distinguished by special cognitive skills or the consistent application of a single scientific method. Rather, they are revealed as practical reasoners, whose judgments are not determined by deductive logic but are pragmatic and contingent. Studies such as those of Lynch (1984) and Knorr-Cetina (1981) have particularly emphasized that scientists use reasoning skills that are entirely comparable with those of people in other everyday situations. All of us engage in pragmatic, contingent judgments that respond to elements in the situations in which we find ourselves and project solutions to the problems with which we are presented.

Second, however, the laboratory site is consequential in differentiating scientists' work from the more widely distributed accomplishments of practical reasoning. All of us make situated judgments, but not all of us do so with the specific resources that the laboratory makes available. Material and human resources are concentrated there, the passage of people and things in and out is carefully regulated, and elaborate practices are inculcated for communicating the results of work there to the wider world. Latour and Woolgar emphasized the importance of "inscriptions" – the visible traces yielded by various kinds of instruments, which can subsequently be represented in written documents to support the plausibility of the authors' statements. These statements acquire greater authority as they are repackaged in different discursive forms (printed papers, review articles, eventually textbooks), becoming progressively stripped of "modalities" – the qualifiers that weaken purportedly absolute factual claims by referring to the circumstances of their origin. Pressing the interpretive claims of their study to the limit, Latour and Woolgar wrote: "It is not simply that phenomena *depend on* certain material instrumentation; rather, the phenomena *are thoroughly constituted by* the material setting of the laboratory. The artificial reality, which participants describe in terms of an objective entity, has in fact been constructed by the use of inscription devices" (1979/1986: 64).

Other analysts have agreed that the focus on laboratory work shows scientists engaged in the creation of an "artificial reality," that is to say, a configuration of the material world that is *real* but is not to be identified with the "nature" that is supposed to exist prior to and independently of human intervention. The philosopher Ian Hacking has talked of experiment as the business of "creating" phenomena, stabilizing them, and making them reproducible. Phenomena, he suggests, should not be thought of as "summer blackberries there just for the picking," but as entities that are made by instrumental engagement with the material world (Hacking 1983: 230). Laboratories, then, are the places where phenomena, or what the French philosopher Gaston Bachelard called "phenomeno-techniques" – phenomena embedded in technical practices – are produced (Bachelard 1980: 61). The special resources of laboratories, both human and material, make them privileged places for the construction of artificial reality.

Both of the fundamental claims of the laboratory ethnographies – that scientific work is practical reasoning in a specific setting, and that human and material resources are constitutive of experimental phenomena – can be traced back to the work of Ludwik Fleck (1896–1961), a Polish/Jewish immunologist and microbiologist, who has emerged posthumously as a long-forgotten pioneer of the sociology of science. Fleck's *Genesis and Development of a Scientific Fact* (1935/1979) was largely neglected when first published, but since its translation into English has belatedly attracted considerable attention (Cohen and Schnelle 1986).

Fleck was insistent that the production of knowledge, even of the "hardest" experimental facts, was a social enterprise and hence an appropriate object of sociological analysis. In one of his more programmatic statements, he declared, "Cognition is the most socially-conditioned activity of man, and knowledge is the paramount social creation" (Fleck 1935/1979: 42). The social character of knowledge is revealed by the circumstances of its production within a specific interactive community (a "thought collective" or *Denkkollectiv*), which sustains a distinctive mode of reasoning (its "thought style" or *Denkstil*). Fleck sometimes wrote of thought collectives as relatively large-scale entities, such as political parties, nations, or all of the practitioners of a scientific discipline; but most of his analysis relied upon an application of the term to much smaller units, the subcultures or research groups encountered in experimental laboratories. Rather than attempting to define the social profiles of such groups from external criteria, he appreciated that their identity and cohesion were bound up with their engagement in common tasks and their possession of shared knowledge. This is one of the respects in which Fleck anticipated the sociology of science that has since taken its lead from Wittgenstein's notion of forms of life. Fleck's richly detailed evocation of his own experimental work eloquently conveyed how knowledge is involved with practical, tacit skills, and how much the inculcation of these skills relies upon concrete exemplars of good practice.

It is in the setting of a specific thought collective that experimental facts arise, according to Fleck. The researcher, he wrote:

looks for that resistance and thought constraint in the face of which he could feel passive. . . . This is the firm ground that he, as representative of the thought collective, continuously seeks. . . . This is how *a fact* arises. *At first there is a signal of resistance in the chaotic initial thinking, then a definite thought constraint, and finally a form to be directly perceived.* A fact always occurs in the context of the history of thought and is always the result of a definite thought style. (Fleck 1935/1979: 94–95; emphasis in original)

Although a fact is thus the result of an experience of passivity in the face of material resistance, it can only be encountered in the context of certain local practices of reasoning, instrumentation, discourse, and so on. Facts are not the results of collective imagination or delusion, but they cannot exist independently of scientists' activity; they are experienced as objective, but they occur only within specific settings of practice. When experimenters seek for facts, according to Fleck, they are seeking for the "passive linkages" that arise from the activity characteristic of their particular thought styles. In summary, "the *fact* thus represents a *stylized signal of resistance in thinking*" (1935/1979: 98).

Fleck realized, however, that to describe experimental facts as products of specific, localized practices raises an important question, one that might be said to hamper all contextual accounts of scientific knowledge: If science is a product of local forms of life, how does it come to have universal application? If experimental knowledge is made with the resources of a particular place, how can it be found to be valid elsewhere? Because of the general importance of this problem, and of the solutions that have been proposed to it, I shall label it "the problem of construction." In a sense, this is the same problem that traditional philosophy of science has tackled under the heading, "the problem of induction," with the assumption that the question is one of legitimating the form of argument that moves from a particular instance of a phenomenon to a general law. Collins's work, however, yielded persuasive evidence that no general theoretical solution to the problem could be found. General rules do not seem sufficient to specify how replication can be accomplished. Fleck suggested instead that the problem demanded a more pragmatic approach – scrutiny of the practices by which the transfer of phenomena from their site of origin was achieved. It would be necessary to investigate how laboratory-made facts are made to hold good outside the laboratory. What work has to be done to make experimental phenomena travel to locations away from where they were constructed? Fleck provided two answers to this question, both of which have subsequently been taken up and developed within constructivist studies. One concerns the mechanisms of communication, the other the transfer of the sustaining culture of thought collectives to new sites.

First, Fleck illuminated the process of transfer of experimental knowledge on what might be called the discursive level. He scrutinized forms of communication within the subculture of the thought collective, and especially across its boundaries. His proposition here was that an important effect occurs in communication between what he called "esoteric" and "exoteric" realms. As facts are translated from the language in which they are represented among specialists to language appropriate for a lay audience, they become consolidated as knowledge. As experts describe their findings to nonexperts, facts are simplified and rendered more dramatic, and the sureness with which they are held is strengthened, even among the experts themselves. As Fleck put it, "Certainty, simplicity, vividness originate in popular knowledge. . . . Therein lies the general epistemological significance of popular science" (1935/1979: 115). The same process occurs in the course of translation from reports in scientific journals to "vademecum" or textbook science. In textbooks, facts are consolidated, simplified, and stripped of reference to the particular circumstances of their origins; they thereby become more certain as knowledge. Even communication between experts in different fields will process facts in the same way. A microscopist, for example, who reports observations of a bacterial culture to a physician who wants help in diagnosis will simplify the findings and make the connection with disease symptoms more direct (113–114). Collins has summarized the process Fleck described here with the slogan "distance lends enchantment." He states that "The degree of certainty which is ascribed to knowledge increases catastrophically as it crosses the core set boundary in both space and time" (Collins 1985: 144–145).

The process of linguistic communication is only part of the story of the extension and consolidation of scientific knowledge. Fleck was certainly *not* suggesting that the certainty ascribed to experimental facts is just a trick of verbal translation. The second condition for the transfer of experimental knowledge away from its place of origin is the motion of the material phenomena themselves. Fleck's claim here was that phenomena are transferred as a package, comprising material and cultural elements together. Laboratory-made phenomena can only be reproduced at another site if the instruments and skills involved in their production are also in place. Fleck's example, described in considerable detail, was August von Wassermann's serological test for syphilis, developed in the first decade of the twentieth century.

Wassermann's procedure provided Fleck with a good example of a constructed experimental phenomenon. The discovery did not simply correlate with a prior theoretical understanding of syphilis, but fundamentally reconfigured the concept of the disease. Along with the bacteriological findings that occurred around the same time, the serological test cut across a confused and contested field of clinical symptoms, therapeutic responses, and experimental results. The previously accepted association of syphilis with the diseases gonorrhea and soft chancre was severed, while such other symptoms as hard chancre and progressive paralysis were included with syphilis as the results of infection by the same bacterial agent, *Spirochaeta pallida* (Fleck 1935/1979: 1–19; cf. Bloor 1983: 34–37). This was, however, a highly negotiated achievement. Bacteriological and serological observations were constantly compared and adjusted with respect to one another; and this continued to be necessary as the Wassermann test was refined. The development of the test itself followed an entirely unanticipated path. Fleck identified a fundamental shift in its understood purpose, from detection of syphilitic agents themselves (antigens) to detection of antibodies formed in infected serum. The first publications by Wassermann and his coworkers indicated that the procedure would be more useful for the former, but subsequently the latter assumed more importance. A further modification of the technique, in which alcohol extracts from healthy organs were substituted for infected ones, could not be explained within the bounds of Wassermann's original immunological understanding of the reaction at all. To Fleck himself, the theoretical explanation of why the test worked still remained unclear (1935/1979: 52–81).

The procedure was nonetheless made to function as an effective test by modifying the techniques used and dispersing them as a package. The sensitivity of the test was deliberately decreased to reduce the number of false positives and enhance its clinical utility. Extensive comparison with clinical observation nonetheless remained necessary to eliminate the many other diseases that could still produce false positives. Application of the test demanded supplies of purified materials and a series of control samples for comparison. Success required training in techniques of quantitative analysis and a somewhat mysterious "experienced eye or serological touch" (Fleck 1935/1979: 53). For Fleck, this successful extension of laboratory-made knowledge was the result of a social process – the widening of the range of a thought style by expansion of a thought collective. He wrote:

> The findings were stabilized and depersonalized. This thought collective made the Wassermann reaction usable and, with the introduction of the alcohol extract, even practical. It standardized the technical process with genuinely social methods, at least by and large, through conferences, the press, ordinances, and legislative measures. (78)

Joseph Rouse has described this process in a way that draws upon Fleck's classic account and the more recent writings of Hacking, Collins, Latour, and others. He describes laboratories as places where "phenomenal microworlds" are created – that is, where material and cultural conditions are manipulated artificially so that experimental phenomena can be manufactured (Rouse 1987: 101). Phenomena are then reproduced outside the laboratory by transferring the conditions prevailing in the "microworlds" to other settings. The apparently universal range of scientific knowledge, taken for granted in traditional philosophical analyses, is viewed as the outcome of laborious processes of "standardization," which involve training in practical skills, mass production of material resources and instruments, and the regulation of units of measurement. Rouse insists that his point is, "not that scientific knowledge has no universality, but rather that what universality it has is an achievement always rooted in local know-how within the specially constructed laboratory setting" (1987: 119). The knowledge produced through this local, practical know-how "is extended outside the laboratory not by generalization to universal laws instantiable elsewhere, but by the adaptation of locally situated practices to new local contexts" (125).

While Rouse has articulated the findings of the laboratory studies in a philosophical idiom, Shapin and Schaffer (1985) have put them to work in historiographical practice. Playing the ethnographic stranger, they closely scrutinized the work in Boyle's laboratory. Like the sociologists, they examined the discursive and technical practices through which Boyle constructed experimental facts: his use of the air pump and complementary rhetorical and social "technologies" to discipline and persuade the audiences that witnessed its results. Boyle's written narratives of his experiments, and the illustrations that accompanied them, were ascribed particular rhetorical importance. As his verbose and declaredly "modest" rhetoric won acceptance for his factual claims, it simultaneously erected a boundary between fact and interpretation, relegating questions of philosophical theory to the limbo of speculative opinion. It is because Hobbes resisted the way in which this barrier had been imposed as fiercely as he rejected the creation of natural knowledge by artificial instruments that his arguments against Boyle are worth following.

But, by following Hobbes's resistance, we might well end up with an enhanced appreciation for what Boyle achieved. Notwithstanding Hobbes's claim that issues of politics and religion were at stake in Boyle's implicit philosophical position, Boyle successfully persuaded many observers that it was only what happened in the air pump that mattered. His demonstration of the "spring of the air" provided a substantial buttress to this delineation of the realm of experimental fact. As Latour has pointed out (1990: 151–152), Shapin and Schaffer have in effect performed a similar maneuver: Boyle made the argument with Hobbes devolve on what happened in the pump; similarly, Shapin and Schaffer focused the attention of the historian onto the technical details of the laboratory. It is in that setting, they insisted, that issues of the social nature of scientific practice must be addressed.

In this way the two authors exploited the potential of the controversy studies for exhibiting the social dimension of the technical contents of science. They also provided an exemplary analysis of the transmission of laboratory-made facts to other locations. They devoted a lengthy chapter (chap. 6) to recounting the labor involved in recreating the culture of Boyle's air-pump experiments at other places in England, and in France, Holland, and Germany in the 1660s. They showed how fragile the experimental culture was that sustained production of the phenomena associated with air pressure, and how it could only be transmitted effectively when literary communication was supplemented by other techniques. Instruments had to be shipped from place to place, individuals traveled to and fro conveying firsthand experience of how to work them, and witnesses were marshaled at each site to give their testimony as to what they saw. It was through such work to extend the material and cultural context in which the phenomenon had been constructed that it gradually came to be reproduced more widely.

We can therefore draw upon Fleck and Shapin and Schaffer for a pivotal lesson of laboratory studies: The local setting where scientific knowledge originates is crucial, but the wider realm beyond its walls can also be viewed as an arena in which knowledge is constructed. While the ethnographic stance might seem to point only toward microstudies of particular laboratories, constructivism need not draw its borders so narrowly. The proclaimed universality of science can be portrayed as a human creation, the result of large-scale extensions of local forms of life. Two strategies have been suggested for investigating this process. First, consider the embeddedness of phenomena in localized technical and cultural practices. Second, scrutinize how the context itself is mobilized to sustain reproduction of the phenomena elsewhere. Look, in other words, for standardization of techniques, instruments, materials, and skills; and consider the effect of combined material, social, and discursive "technologies."

Many of these points were developed, and wittily expressed, in Bruno Latour's *Science in Action* (1987). In many respects, Latour's program was built upon the foundations of the sociology of scientific knowledge, with which he was intimately familiar. However, as Shapin pointed out in an incisive review (1988a), although Latour's arguments can be assimilated in certain respects to the previous themes of SSK, his program as a whole was presented as an alternative, which distanced itself radically from the sociological project as it had been conceived. In subsequent writings, Latour went on to amplify his account of these differences. He called into question the whole aim of a sociological explanation of scientific practice, suggesting that neither the goal of explanation as such nor the categories usually resorted to by sociologists should be maintained (Latour 1988b, 1990, 1993). What was needed, he argued, was "one more turn after the social turn" in science studies (Latour 1992). Because of the originality of his approach, Latour's work both contributed new resources and posed new problems for historical study of the sciences, over and above those deriving from sociological studies in general. It therefore seems necessary to discuss at some length Latour's perspective, and some of the debate it has provoked, by way of concluding this survey of constructivist theory and its implicit models of the social realm.

To follow, initially, the route of assimilating Latour to previous sociological studies, we can note what he has offered by way of solving "the problem of construction." Indeed, his work contributes significantly to an understanding of the processes by which experimental knowledge is transferred from its point of origin. The discursive and material mechanisms of communication, initially identified by Fleck, are both described vividly in *Science in Action*. The first chapter is devoted to the rhetoric of scientific writings. There, Latour decodes the dense undergrowth of references that usually accompanies a scientific paper as a means of buttressing the author's statements with the authority of predecessors and colleagues. Scientific rhetoric is a stylized way of enhancing the credibility of written assertions by raising the cost of dissent. To resist the claims of a well-documented paper is to pick an argument with all of the authorities it cites. The most "technical" literature is thus, Latour points out, the most "social," in the sense that it mobilizes the largest number of authorities. Its factual status is further increased to the degree that it is cited by subsequent writers, whose reinscription of statements enhances their factuality by distancing them from the particular circumstances in which they originated. These are the discursive means by which "facts" – at least in their textual manifestation – are constructed. In a sentence that could almost have been written by Fleck himself, Latour sums up: "a fact is what is collectively stabilised from the midst of controversies when the activity of later papers does not consist only of criticism or deformation but also of confirmation" (1987: 42).

Latour quickly goes on to dispel any impression that he is suggesting that scientific discourse is purely literary or fictitious. If a reader continues to refuse to accept a written claim, one can resort to the laboratory instruments that produced the original "inscriptions" represented in the text. Whatever rhetorical tools it might share with other forms of writing, science also utilizes crucial nondiscursive resources. Latour's analysis of how these material resources are used nonetheless took much the same form as his account of the rhetorical momentum of scientific literature. Instruments are "black-boxed," – that is, accepted as working machines that exhibit phenomena of nature – and are then passed from user to user, enhancing their authority as they move away from their place of origin. The thermometer, for example, which in the seventeenth century was a problematic and disputed object of research, came gradually to be used in a taken-for-granted way as an instrument for investigating chemical and meteorological phenomena. By the late eighteenth century, it had become accepted as an unproblematic resource by means of which other phenomena (such as heat exchanges in chemical reactions) could be explored. Black-boxed machines, like the written claims that gain in factual status as they pass from text to text, become more effective as they are used by more and more people. Latour writes: "the black box moves in space and becomes durable in time only through the actions of many people; if there is no one to take it up, it stops and falls apart however many people may have taken it up for however long before" (1987: 137).

In Latour's view, facts and machines are transmitted in the same manner, passed down the chains of linked persons and things that he calls "networks." Networks are fundamental to the motion from local knowledge in laboratories to worldwide "technoscience" (Latour's chosen term for an inseparable merging of science and technology). They comprise Latour's solution to the problem of construction. In many respects, Latour's vision of a world encompassed by ever larger technological and scientific networks is conformable with that depicted by other sociologists of science. By extension of these networks, the world is seen to be reshaped to resemble the special settings in which experimental artifacts are made. Hence, facts and machines come to be able to survive in the world beyond the laboratory walls.

It is when we consider, in more detail, Latour's account of how networks are built that the differences between his perspective and those of his sociological colleagues begin to become apparent. For Latour, an understanding of technoscientific networks requires that we recognize the agency of nonhuman, as well as human, beings. This calls into question traditional understandings not only of science but also of human society.

Latour's networks are heterogeneous linkages of people and things – associations of human and nonhuman entities, to which he gives the common name "actants," borrowed from semiotics. Thus, we are told that Louis Pasteur had to learn to control both bacterial cultures (which required an appropriate nutrient medium to grow in his Paris laboratory) and provincial farmers concerned about outbreaks of anthrax (who were persuaded by publicity and carefully staged demonstrations that Pasteur could help them). The success of the Bell Company in constructing the first transcontinental telephone line in 1914 is said to have depended upon linking together physicists trained in Robert A. Millikan's Chicago laboratory and electrons that amplified the signals passing through the new electronic repeaters (Latour 1987: 123–127; cf. Latour 1988a). The large-scale systems in which scientific and technological artifacts come to be extended across time and space are composed by "enrolling" such human and nonhuman elements in heterogeneous associations, Latour argues.

In a recent defense of their "actor-network" approach, Latour and his collaborator Michel Callon have explained that they are interested in mapping the construction of these heterogeneous networks of human and nonhuman "actants" because this offers a way of following (though not exactly explaining) the passage from controversial to accepted knowledge (Callon and Latour 1992). They criticize the alternative approach of SSK, represented by the controversy studies of Harry Collins. Collins, they note, typically argues that, since "nature" is not determinative of scientists' decisions about what to accept as knowledge, then "society" must be. He invokes networks of trust and credibility to explain how this societal constraint comes to bear upon the judgments that scientists make. The difficulty, according to Callon and Latour, is that purely social relations seem too weak to do the job. To claim that what determines the outcome of a controversy is a decision to accept one set of evidence rather than another – a decision based upon the social ties among individuals in the community – is to rest a purported explanation upon ground far too shaky to support it. This line of argument invites dismissal for dealing in "merely social" explanations, Callon and Latour assert (1992: 352–356).

More positively, the advocates of the actor-network propose that their approach should be viewed as an extension of the underlying principles of SSK rather than a contradiction of them. They seek to capture the ways in which the ontological status of entities is itself an open question during the course of controversies. Decisions as to what is a real object and what is an effect of human error, what is an artifact of instrumentation and what is a phenomenon revealed by refined human skills, are made only upon the resolution of controversies. Entities are assigned to the human or nonhuman realm as debates are settled. Hence, in an extension of the postulate of symmetry (which was foundational for SSK), Callon and Latour propose that the analyst has to keep an open mind about where things belong in these ontological categories: "If engineers as well as scientists are crisscrossing the very boundaries that sociologists claim cannot be passed over, we prefer to abandon the sociologists and to follow our informants" (1992: 361).

There are many issues raised by the actor-network approach to the problem of construction, and, as one might expect, it has provoked considerable debate. I shall explore just two areas of this debate here: first, the question of the form and status of the analysis that is being proposed. At times, Callon and Latour seem to be advancing their actor-network theory as a kind of semiotics, in other words, as an account of how the actants they identify function as signifiers in a discursive field. In the discourse of scientists and engineers, effects may be assigned to humans or to nonhumans, and the two classes of entities may exchange signifying roles. Humans may be substituted for nonhumans or vice versa. Much of Latour's analysis of Pasteur's work is couched in these terms, using the texts authored by "Pasteur," in which textual signifiers like "anthrax baccilli," "sheep farmers," and so on, are manipulated, redistributed, and assembled in new relationships (Latour 1988a). But, one cannot help asking, how does this relate to actions in the real world? Are nonhumans to be seen as completely substitutable for humans, and vice versa, in actual practice? There are certainly many times when Latour seems to be going well beyond semiotics to advance an *ontology* of a radical new kind, in which nonhuman entities are ascribed an equal degree of agency with humans (cf. Lynch 1993: 107–113).

Most commentators have found this quite implausible. The conflation of semiotics with ontology seems like a failure to distinguish between reinscribing the accounts of scientists themselves and giving a detached sociological analysis (Gingras 1995). It is one thing to say that human social life is substantially shaped by the participation of nonhumans, but it is quite another to draw the conclusion that no distinctions can be made between the two categories. Many contemporary developments, such as the increasingly pervasive use of computers, artificial body parts, genetic engineering, and the concern for the legal rights of animals, indicate the complex interpenetration of nonhuman entities and the human world. These are the kinds of developments Donna Haraway has pointed to, in her essay on the contemporary prevalence of "cyborgs" – the name she gives to these heterogeneous associations of human and nonhuman entities (1991). Insofar as they are simply pointing to the ways in which social life has been affected by these associations, Callon and Latour have a strong point when they say, "There is no thinkable social life without the participation – in all the meanings of the word – of nonhumans, and especially machines and artifacts. Without them we would live like baboons" (1992: 359).

Participation in social life does not, however, imply a degree of agency equal to that of humans. It would seem perverse to deny that human life is lived in interaction with the material world, and that material things can sometimes act to thwart human desires. Recent experience of technological change has driven home the lesson that functions can indeed be redistributed between humans and nonhumans, for example, from skilled workers to machines. But this does not mean that analysts can afford to obliterate distinctions between human and nonhuman action. Collins (1990), for example, has argued that close study of how computers are treated in society shows that they should not be assigned agency in anything like the human sense: In many respects, they depend for their acceptance upon humans learning new skills for dealing with them. Of course, it also seems clear that computers have participated in teaching us how to interact with them, and that they do enable human skills to be restructured in a way that would not be possible in their absence. If machines cannot be said to have agency on a level with humans, then, they are at least capable of redirecting human agency and are not entirely subject to it. To that degree, Callon and Latour have made a strong point.

There are profound philosophical questions at stake here, which have surfaced in recent discussions of Artificial Intelligence, for example. But, for historians, a more immediate issue is how historical accounts are to be framed in light of the actor-network approach. This is the second aspect of the debate that I want to explore: the question of historical narrative. In a critical appraisal of Latour and Callon's work, Collins and his coauthor Steven Yearley have teased apart the language of some of their exemplary case studies (Collins and Yearley 1992a, 1992b). Callon's account of the network constructed by marine biologists in the St. Brieuc Bay in Brittany is subjected to a particularly close reading (Callon 1986). Callon's claim was that the network he was tracing could only be adequately understood if the scientists involved were seen to be "negotiating" equally with their colleagues, the fishermen working the bay, and the scallops they wanted to collect. They persuaded their colleagues that the collectors they employed anchored the scallops to a "significant" degree; they persuaded the fishermen that their experiments were not a threat to fishing and hence should be left alone; and they persuaded the scallops, by "negotiating" with them, to attach themselves to the collectors. Collins and Yearley argue, however, that there is a break with the fundamental postulate of symmetry in the way Callon has told the story. By ascribing a role to the scallops in resolving the controversy about the degree to which they attached to the collectors, Callon has made a judgment in favor of one side of the debate. Far from sustaining a symmetrical neutrality, the analyst has intervened to assign to the scallops a decisive role.

The problem with such a departure from neutrality, according to Collins and Yearley, is that it reverts to a realist style of narrative with which the symmetry postulate was designed to break. Callon's descriptions can very easily be translated into whiggish realist stories, in which the outcome of the controversy was always inherent in a reality that had only to be revealed. Collins and Yearley show how simple such a translation is: Only a few items of vocabulary have to be changed (1992a: 315–316n.). And Yves Gingras has amplified the point with a similar reading of other narratives composed by members of the actor-network school, which offer a kind of thinly disguised "neorealism" (1995). Thus, Collins and Yearley have plausibly argued that there is a serious problem with the particular way in which Latour and Callon have introduced nonhuman actors into their narratives. They have endorsed the stories of the "winners" of scientific disputes, and thus made it impossible to account for the process of resolution of the disputes as other than the disclosure of a preexisting reality.

Schaffer has shown how this may affect the framing of historical accounts, in a review of Latour's monograph on Pasteur (Schaffer 1991, cf. Latour 1988a). Latour's "hylozoism" – his ascription of living agency to nonliving things – is shown by Schaffer to have led to an unbalanced historical narrative. Schaffer notes that Latour invokes the dubious figure of an "ideal reader" to assign roles to the various semiotic entities he describes. In doing so, he basically reinscribes Pasteur's own semiotic maneuvers. But, Schaffer proposes, an account quite different from Latour's would be necessary if one were to take seriously the viewpoint of Pasteur's long-standing rival, Robert Koch. While Latour portrays Pasteur as successfully enrolling microbes in his network, Koch would not have agreed. So long as the humans are engaged in a debate over what the properties of the nonhuman actors *are*, Schaffer insists, the postulate of symmetry requires that the analyst remain neutral. An account of the debate should not ascribe to nonhumans the ability to decide the issue, because the author can only do so by identifying himself with the winning side. After all, the microbes did not tell their side of the story; it was Pasteur who was nominated their "spokesperson."

The suggestion of these critiques is that Latour and Callon have located nonhumans in the networks they describe according to the subsequent outcomes of disputes that were still raging while the networks were being built. Their retrospective narratives of the enrollment of nonhumans obliterate the openness and uncertainty that always surrounds science in the making, by reaching forward to the situation after the debates have been resolved. What is needed, this suggests, is more study of how scientific practice *unfolds* in time, and how it *enfolds* the material world with human society. An understanding of how nonhuman entities are incorporated in, and help to transform, human social life will require careful reconstruction of incidents of practical engagement with the material world. The challenge is to develop historically sensitive narratives, in which the openness and the resistance experienced by human actors in the course of dealing with the world would be adequately accounted for.

Historians trying to meet this challenge may continue to draw upon the resources provided by sociological reflection. Two recent perspectives seem especially pertinent to the task of going beyond the debate between the actor-network and its critics: first, the notion of "boundary objects" developed by Susan Leigh Star and James R. Griesemer (1989) and by Joan H. Fujimura (1992); and, second, Andrew Pickering's "mangle of practice" (1995a). Both perspectives offer ways to think about the mutual interpenetration of human action and the world of nonhuman things. The first tends to emphasize the dimension of social groups of practitioners; the second, that of the temporality of practice.

The notion of "boundary objects" has been applied by Star and Griesemer to an analysis of the creation of the Museum of Vertebrate Zoology at the University of California in the early twentieth century, and by Fujimura to recent research on cancer genes or "oncogenes." Boundary objects are things that link together different social groups, who may view and use them in quite different ways. Collections of vertebrate specimens, for example, may mean quite different things to professional zoologists, university administrators, and amateur collectors. Boundary objects could also include images or representations, artifacts or naturally occurring objects, elements of the physical environment or the inscriptions of instruments. These objects exist in distinct "social worlds," to the extent that they are understood differently and used to advance different aims by the communities involved. Rather than focusing on what divides these groups, however, the analysis emphasizes what they *share*, notwithstanding their different outlooks. Star and Griesemer assert that the objects in question must have a degree of robustness in order to move between and hold together the separate social realms in which they are used:

> Boundary objects are objects which are both plastic enough to adapt to local needs and the constraints of the several parties employing them, yet robust enough to maintain a common identity across sites. . . . They have different meanings in different social worlds but their structure is common enough to more than one world to make them recognizable, a means of translation. (Star and Griesemer 1989: 393)

This kind of analysis portrays a process of making knowledge through embedding elements of the material world in human practice. Stabilization is achieved, not through a straightforward consensus, nor through the building of a network outward from a single source, but through the trading of objects across boundaries between different social realms. Facts are constructed, on this view, by being passed between social domains that retain a degree of autonomy but are nonetheless linked by their shared use of the boundary objects. A recent application of this perspective to a historical case study has been provided by Anne Secord (1994), who demonstrates how the trading of such objects as botanical specimens and schemes of classification linked the separate realms of gentlemanly and working-class naturalists in early-nineteenth-century England.

Pickering's "mangle of practice" highlights the dimension of time in relation to scientific practice. Cutting a line between Latour and Collins, Pickering declares that he is willing to ascribe a degree of agency, but not of intentionality, to nonhumans. He nonetheless distances himself from retrospective realism by insisting that material agency is "emergent in time": It is encountered in the course of practice, in which human and nonhuman action interactively engage one another toward the end of a stable configuration. From the human point of view, experimental inquiry takes the form of a "dialectic of resistance and accommodation," until the achievement of stability results in isolation of a phenomenal "reality" framed by instrumental technique (Pickering 1995a: 9–27). Pickering's is also, however, a "posthumanist" theory, in the sense that he does not see the process of inquiry as directed by stable human aims. The intentions and interests of the investigators, too, may be transformed by passage through the "mangle of practice."

This is an intriguing model, which acknowledges the oscillation between action and resistance in experimental practice, referred to (for example) by Fleck in the passage quoted earlier. It also accounts for the mutual entanglement of the objects of investigation with the instrumental tools used by human investigators, the two sides only being distinguishable when a stable configuration has been accomplished. The emergence of a clear and stable distinction between the phenomena and their instrumental framework is always liable to be projected back with the advantage of hindsight. Hence the special challenges posed for the history of scientific practice, insofar as it seeks to overcome the perspective of hindsight and recapture the original experience of experimental inquiry.

Because of its engagement with the issue of temporality, Pickering's vision seems an important one for historians to ponder. I shall return to consider its implications for narrative reconstructions of scientific practice at the end of this book. It stands as an encouraging sign of the continuing value of the relationship between history of science and constructivist sociology. The traffic between the two disciplines has been in both directions: As sociological debates about nonhuman agency have raised issues for historical narrative, so narrative has been seen as an arena in which solutions to the theoretical problems may be worked out. Pickering has contributed to this himself with the series of case studies that his book presents. It is also striking that the two sides in the debate between Callon and Latour and Collins and Yearley both pay the tribute to the historians Shapin and Schaffer of enlisting their work in support of their claims. Perhaps this is because the historians' rich narrative succeeds in conveying the experience of the openness of scientific practice and of its engagement with material objects. Boyle's air pump was evidently a real object; provided with the right cultural setting, it could convey the phenomena of the spring of the air to new locations. But the pump's sealant may or may not have been really leak-proof; the authors say only that Boyle stipulated it was but that others disputed this. Shapin and Schaffer generously populate the world they describe with certain material entities, but, insofar as other entities are subject to debate, they refuse to meet the expectations of readers who want to know what was "really going on."

At this point, where we find the sociologists pointing to historical practice as exemplary, we can appropriately conclude our review of constructivism. Issues of the social role of material artifacts, the degree to which they are subject to human agency and capable of redirecting it, will need to be resolved in narrative reconstructions of scientific investigation. It is to the historians, then, and to the implications of constructivism for their practice, that we shall now turn.

2

Identity and Discipline

But also when I am active *scientifically*, etc., – when I am engaged in activity
which I can seldom perform in direct community with others – then I am
social, because I am active as a *man*. Not only is the material of my activity
given to me as a social product (as is even the language in which the
thinker is active): my *own* existence *is* social activity, and therefore that
which I make of myself, I make of myself for society and with the consciousness of myself as a social being.

Karl Marx, *Economic and Philosophic Manuscripts* (1964: 137)

THE MAKING OF A SOCIAL IDENTITY

One theme of the previous chapter might be stated as follows: Constructivist studies of science have helped us to see how understandings of
"nature" are products of human labor with the resources that local cultures make available; but they have also pointed to a need to revise our
ideas of "society." In the course of the development of the constructivist
approach, new ideas of the social context of scientific practice have been
explored. Analysts who have applied Wittgenstein's notion of "forms of
life" have portrayed social formations ("paradigms" or "core sets," for
example) that are defined by particular configurations of scientific practice. These are relatively fluid entities, not directly conformable to the
institutions or professional communities that sociologists of science have
traditionally analyzed, and somewhat autonomous from society at large.
Latour has made the break with traditional sociology much more decisively, arguing that the practitioners of science and technology reconstruct their social world as they work to build up their picture of nature.
Not all are willing to accept the conclusions Latour derives from this –
that the usual social ontologies should be dispensed with and "one more
turn after the social turn" taken – but it is clear that some new approaches to mapping the social profile of science are required.

In this chapter, we consider some of the implications of this for historical studies of the scientific community and of the place of individuals
within it. What does the constructivist perspective have to offer in analyzing those topics that have traditionally been the preserve of sociolo-

gists of science and social historians: the formation of scientific careers
and disciplines, professionalization, and the creation of scientific institutions? What kind of description might be emerging of the processes
by which scientific practice remakes its social environment as it produces
new knowledge? I shall answer these questions with special reference to
research on two critical periods in the history of science. First, the early-
modern period, when experimental science emerged in Europe, practised
by groups of natural philosophers, in settings such as academies and
universities that remain important to this day. And second, the period
including the end of the eighteenth and the early nineteenth centuries,
sometimes called "the second scientific revolution," when many of the
disciplines and institutions of science assumed a recognizably modern
form.

I begin by defining the constructivist perspective more sharply by
comparison with an alternative. Much social history of science produced
in the 1960s and 1970s reflected the influence of Robert K. Merton. Merton, who taught for several decades at Columbia University, achieved
recognition as the dean of American sociology of science and also sustained a long-standing interest in its history. His historiographical approach was premised upon a firm demarcation between the cognitive
content of science and its social context, which he originally developed
in the 1930s under the influence of his teachers at Harvard, the historian
George Sarton and the sociologist Pitirim Sorokin (Shapin 1992). By
erecting this boundary between what was "internal" and what was "external" to science, Merton restricted the degree to which particular scientific developments could be ascribed to social causes. He wrote:
"Specific discoveries and inventions belong to the internal history of science and are largely independent of factors other than the purely scientific" (Merton 1938/1970: 75; quoted in Cole 1992: 3). His pupil
Bernard Barber subsequently formalized the distinction between internal
and external "factors" in scientific change: "The internal factors include
those changes which occurred within science and rational thought generally; the external include a variety of social factors" (quoted in Shapin
1992: 340).

The distinction between internal and external factors has since been
used predominantly by philosophers and historians who have been concerned to keep science pure from any taint of contact with social forces.
They have set up camp on the "internal" side of the Mertonian divide
and fortified their position against the incursion of any "external" causes
liable to impinge upon the progress of science. Although the dichotomy
clearly offers itself to this kind of analysis, Merton himself did not originally use it for that purpose. For him, there were crucial mediating factors that permitted a connection, albeit a weak one, between the external
and internal realms. These were the socially embedded values that sus-

tained the pursuit of science in a particular context. The degree to which such values were present in the surrounding culture, or entrenched in scientific institutions, could affect the rate of scientific progress and even modify its direction. There was, however, no question of the specific products of inquiry – the cognitive core of science – being affected by external influences. To this extent, Merton was indeed fencing off the ground of idealist historiography to protect it from sociological trespassers.

Merton's widely discussed thesis about the influence of Puritanism on science in seventeenth-century England aimed to show how widespread cultural values could encourage the pursuit of natural knowledge prior to the existence of independent scientific institutions (Merton 1938/1970; Abraham 1983; Shapin 1988c). In the period since the seventeenth century, he envisioned a gradual process of increasing autonomy of those institutions, whereby they came to internalize the ethical principles fundamental to scientific progress. This is what Thomas Gieryn calls Merton's "postulate of institutional differentiation," according to which the processes of evaluation and reward came to be routinely administered within the scientific community (Gieryn 1988).

The moral principles governing scientific inquiry were the so-called norms, four of which Merton claimed could be identified in "the moral consensus of scientists . . . and in moral indignation directed toward contraventions" (Merton 1942/1973: 269). First is "universalism," the injunction that claims to truth be assessed independently of the attributes of their proponents – "The Haber process cannot be invalidated by a Nuremberg decree nor can an Anglophobe repeal the law of gravitation," as Merton put it (270). Second is "communism," by which he meant the disavowal of secrecy or private-property rights in knowledge. Discoverers are rewarded only by honor and commemoration within the scientific community, not by retaining private rights over their discoveries. Third, "disinterestedness" is enjoined as a further check upon self-aggrandizing ambition or deliberate fraud. Finally, "organized skepticism" is a mental habit expected of scientists, even though it may give rise to conflicts with the upholders of cultural and religious traditions.

Taken together, the four norms constitute an "ethos," which must hold sway if science is to flourish or, indeed, in order for anyone to occupy the social role of "scientist." The historical origins of the ethos might be sought in widely distributed cultural values but, once established, the norms are sustained through time by a succession of individuals who learn to act as scientists by becoming members of the appropriate institutions. As Norman Storer explained, in his introduction to Merton's collected papers, individuals' behavior comes to be shaped by the norms as they are socialized as members of the scientific community:

Norms of this sort are associated primarily with a social role, so that even when they have been internalized by individuals, they come into play primarily in those situations in which the social role is being performed and socially supported. When scientists are aware that their colleagues are oriented to these same norms – and know that these provide effective and legitimate rules for interaction in "routine" scientific situations – their behavior is the more likely to accord with them. (Storer in Merton 1973: xix)

The Mertonian model gives institutions a central, but strangely occluded, role in the sociology of science. According to the differentiation postulate, institutions arise when they mirror a broader cultural acceptance of the scientific ethos. In seventeenth-century England, for example, the Royal Society was simply a conduit for transmitting prevalent cultural values – those characteristic of Puritanism, in Merton's view. Thereafter, they become vehicles for imparting the norms to initiates, hence sustaining scientific practice over time; they induct individuals into the social role of "scientists." But they remain curiously insubstantial entities. The particular local features of particular organizations, such as academies or universities, do not determine how science is practised or what its products are in those locations. Instead, Mertonian sociologists have tended to direct their attention to the general conditions for maintenance of the ethos, such as the reward system or the structure of peer review. In effect, they treat the context of science as those circumstances which allow it to proceed without being hampered by any particular social constraints. Although dysfunctional organizations can hinder growth, properly functioning institutions are simply channels to convey the fertilizing values that irrigate the field of science.

Mertonian sociologists discount the possibility of any influence by the local setting upon the contents of science because they see institutions as enforcing values that are not local but universal. Their definition of institutions also contributes to an inability to focus upon the material characteristics of particular sites or the distribution of authority among those who inhabit them. As Gary Abraham has noted, Merton was inclined to a view of institutions as "persistent forms of conduct that embody cultural values," rather than as material settings, local groups, or formal organizations (Abraham 1983: 374). Mertonian sociology thus offers few resources for analyzing the specific features of the locations and small-scale groups within which natural knowledge is produced.

To see what kind of history results from such an orientation, we can consider Joseph Ben-David's *The Scientist's Role in Society* (1971/1984). Ben-David was not directly a member of Merton's school, but the assumptions underlying his history closely parallel those of Merton himself. For example, Ben-David takes it for granted that he can unproblematically identify something designated "science" throughout the period he describes, from ancient Greece to the twentieth century. In this

sense, his book is similar in conception to the philosophical macrohistories (the "big-picture" accounts) of science that we saw had undergone
eclipse since Kuhn; it buys its purported comprehensiveness at the cost
of positing a transhistorical notion of what science is.

Ben-David's work is Mertonian also in that its basic theme is the process by which science achieved autonomy from external influences. In
other words, and somewhat paradoxically, it is a sociological history that
aims to show how scientists attained independence from social forces.
Three stages of this process are to be identified: the creation of a "social
role" for the scientist; the achievement of "intellectual autonomy" for
science; and the construction of organizational autonomy in institutions
devoted to the subject. As with Merton's approach, then, institutions
assume a position in the analysis that is secondary to the creation of the
scientific role with its sustaining system of values.

In the introduction to the second edition of his book, Ben-David carefully differentiates his project from "the sociology of the contents of scientific knowledge" (1971/1984: ix). The latter is an enterprise with very
limited possibilities, in his view, since sociological influences upon the
actual products of scientific inquiry can only be tenuous and sporadic.
In making this assertion, he yokes a Mertonian "internal"/"external"
distinction to an equally arguable contrast between history and sociology:

> The argument . . . does not deny that research and discovery are on many
> occasions influenced by conditions external to science. It only asserts that
> whereas the influence of the internal disciplinary traditions is permanent
> and ubiquitous, since these traditions more or less determine what can be
> done in science at any given time, the external influences are ephemeral
> and random. Therefore, . . . the external influences are proper subject mat-
> ter for historical inquiry, which is traditionally concerned with one-time
> events, but not for sociology, which is interested in regular, 'systematic'
> relationships, and not in one-time occurrences. (Ben-David 1971/1984: xxii)

As an alternative, therefore, Ben-David proposes "an institutional sociology of scientific activity" (14), in which the crucial category will be
the "role" of the scientist, and the investigation will be devoted to tracing
its evolution. He is aware, of course, that the term "scientist" is an invention of the early nineteenth century, but he insists that the social role
designated by that word was first established in mid-seventeenth-
century England. There, the Royal Society first institutionalized the goal
of disinterested empirical research and secured enough social acceptance
to make its program sustainable. Prior to that, the vocation of "scientist"
was held as a personal goal by a few individuals – it is, in fact, something
of a transcendental ideal, common to "a small minority of people in
every society" (xix) – but the personal aim was not translated into a
socially recognized occupation. Ben-David devotes a chapter to consid-

ering why this did *not* happen in ancient Greece, despite the substantial contributions made by certain individuals to scientific progress.

From the historian's point of view, much of Ben-David's conceptualization can be criticized as inherently whiggish. The Greeks are seen as straining forward to attain the social recognition that will only later surround what they are already doing. The seventeenth century awaits the coining of the word that will name the social role it has created. Many questions about what these people themselves thought they were doing, and how it was understood in their cultures, are unanswered within such a teleological model. And teleology also characterizes Ben-David's narrative of the development of early-modern scientific institutions. For him, the main problem posed by this period is, "what made certain men in seventeenth century Europe view themselves for the first time in history as scientists?" (1971/1984: 45). Various progressive trends are educed – a degree of secular independence in the universities, the Renaissance flourishing of the mechanical arts, the foundation of the first intellectual academies in Italy and their translation to Northern Europe – all of them aimed toward culmination in the Royal Society. Even the Protestant Reformation is slotted into its place in this targeted narrative: Readers are told that the Catholic persecution of Galileo was exploited for propaganda purposes by "perhaps the earliest manifestation of an active scientific lobby in Protestant Europe" (72).

In the Royal Society it is, unsurprisingly, the Mertonian norms of scientific behavior that Ben-David sees being put into operation. These norms, including disinterestedness and universalism, were "part of the official program" (76) of the organization, expressed in such works as Thomas Sprat's *History of the Royal Society* (1667/1958). The wider social landscape of seventeenth-century England has little importance beyond providing a receptive climate for these ideals. Ben-David points out that an atmosphere of religious pluralism and social change was tolerant of the free communication of ideas and willing to adapt to new technology. The role of society was thus to grant autonomy to "a scientific community that could set its own goals relatively independently from nonscientific affairs" (Ben-David 1971/1984: 43). The Mertonian notion of autonomy, which is in play here, assumes that a boundary between internal and external realms is a straightforward product of the existence of a self-identified scientific community. Since there is no essential problem about the definition of "science" in the Mertonian approach, it follows that, once people decide to be "scientists," they do best in situations where they are left to get on with it.

This happy result was achieved with the formation of an institution that required its members to adhere to the experimental method. Thereby, in Ben-David's view, theoretical controversy was averted and community solidarity enhanced:

[Empirical science] produced innovations that contained their own incontrovertible proofs and made all philosophical controversy unnecessary. And not only was it a way to innovation, but also to social peace, as it made possible agreement concerning research procedures to specific problems without requiring agreement on anything else. . . . Without an agreement on the experimental method, . . . an autonomous scientific community could never have arisen. . . . By sticking to empirically verified facts (preferably by controlled experiment), the method enabled its practitioners to feel like members of the same "community," even in the absence of a commonly accepted theory. (1971/1984: 73–74)

Superficially, these remarks sound a kind of pre-echo of Shapin and Schaffer's discussion of the formation of experimental culture in *Leviathan and the Air-Pump* (1985). Those authors agree that Robert Boyle's procedures for making experimental "facts" in the Royal Society served to enhance the social cohesion of the group by ruling out metaphysical speculation. They locate the basis of consensus in an acceptance of certain "technologies" of fact creation and the demarcation between a fact and an opinion that those practices implied. But there are also very substantial differences between the approach of Ben-David and that of Shapin and Schaffer, and consideration of these can help us to see how a constructivist perspective departs from the Mertonian tradition of sociology.

One evident difference is that Ben-David, from Shapin and Schaffer's point of view, assumes far too readily that the value of the empirical method was unproblematically perceived – as if everyone was glad to renounce philosophical argument and recognize the authority of experiment. Of course, by paying serious attention to the views of Thomas Hobbes, Shapin and Schaffer show that this was not so (1985: 80–109). Hobbes argued that philosophical doctrine could only be validated by the proven methods of argument and demonstration; manipulation of instruments could not, for him, produce any natural knowledge with the required degree of certainty. So, by using the method of controversy studies, Shapin and Schaffer examine the construction of experimental science without surrendering to the seductive pull of teleology. Rather than taking "scientific method" as a predefined given, and attempting to trace its history back to its roots, they tell their story from a point at which Boyle and Hobbes were engaged in a dispute over fundamental principles, when the outcome was not at all preordained. The implication is that a rigorously historical account of the creation of experimental practice, and of the institutions in which it came to be entrenched, has to pay attention to conflicts between what might have been radically divergent visions of science.

Another characteristic of this approach is a rigorous deconstruction of the Mertonian internal/external distinction. This is not to be identified with the facile gestures toward transcending the distinction, which, as

Shapin (1992) has pointed out, practically everyone has been making since the dichotomy was first announced. The problem is not simply to find ways of linking what is taken to be part of science to what is taken to be outside it, but to scrutinize the ways in which the boundary has been drawn historically and traffic across it managed. Accordingly, we are shown Boyle drawing a line between experimentally produced "facts" and metaphysical "opinions," and supporting the demarcation with techniques to manipulate his apparatus, to frame its verbal description, and to discipline its audience. These techniques are, of course, rhetorical and political; but their effect is to isolate a realm of experimental practice that will be declaredly free from rhetoric and politics. To ignore this achievement would be to neglect what is distinctive about experimental practice; but to overlook the cultural tools with which it was accomplished would be to fail in the duty of historical understanding. The key is to regard the boundary between inside and outside as a constructed entity requiring explanation rather than as a given. As Shapin and Schaffer put it:

> The language that transports politics outside of science is precisely what we need to understand and explain. We find ourselves standing against much current sentiment in the history of science that holds that we should have less talk of the "insides" and "outsides" of science, that we should have transcended such outmoded categories. Far from it; we have not yet begun to understand the issues involved. We still need to understand how such boundary-conventions developed: how, as a matter of historical record, scientific actors allocated items with respect to *their* boundaries (not ours), and how, as a matter of record, they behaved with respect to the items thus allocated. Nor should we take any one system of boundaries as belonging self-evidently to the thing that is called "science." (1985: 342)

In a similar way, the constructivist outlook would tend to look upon the Mertonian norms as an object of historical inquiry rather than an unproblematic resource. While basic values and principles of method might be agreed upon in situations in which consensus exists, the controversy studies have shown how local and transient such achievements of consensus can be. Disputes over matters of fact can be seen to spill over into questions of method, propriety, and competence. In such a situation, principles of "scientific method" may not suffice to ground a resolution of the debate. Although the participants may all regard themselves as scientists, they cannot agree about what the relevant methodological criteria are or how they should be applied to the points in question. Instead, a range of methodological and moral stipulations are articulated, which diverge in their implications or even contradict one another, and which cannot be applied uncontentiously to resolve the issue. These stipulations exist in a complex, context-dependent relation to the statements of fact that they supposedly validate; they should not be regarded as having priority over factual claims, and hence cannot serve as the basis of historical explanation (Schuster and Yeo 1986).

The deconstruction of the Mertonian norms has clear implications for our understanding of institutions. If institutions are not channels for distributing universal values, then they presumably do not provide an off-the-peg identity for those who work inside them. Rather, institutions themselves should be seen as constructs – the outcomes of interactions and negotiations among those who participate in them, as well as of larger-scale forces affecting society as a whole. The constructivist focus is on the local institutional setting of science – the academy, court, university, laboratory, or lecture theater – and on the particularities of practice that characterize it, but organizations are not regarded as inflexible determinants of individual behavior. Individuals may be expected to show a variety of different understandings of their role within a specific institution, and a range of expectations for its purpose. They will probably seek to advance various personal aims while discharging whatever obligations the situation lays upon them. As Gieryn puts it: "Institutions are social constructions in that their definitions, relationships, values, and goals are negotiated by ordinary people in ordinary settings. . . . The institution is available for multiple and not always consistent descriptions and explanations" (1988: 588–589).

To see what consequences follow from adopting such an approach, we can consider recent studies of the early Royal Society. Although not all of these have been informed by an explicitly constructivist orientation, their general drift has been broadly compatible with the perspective I have just outlined. Rather than portraying the institution as unified by a consensus acceptance of certain values, as Merton and Ben-David suggested, recent research has been directed at rigorous contextual interpretation of the documents that have been read as embodying such values. The implication is that methodological stipulations cannot be read straightforwardly as norms. At the same time, various individuals within the Royal Society have been shown to have been pursuing their own agendas and negotiating their own status within the organization. At moments when disputes erupted, communal norms seem to have exerted little restraint. Instead, they served as rhetorical weapons, apparently adaptable to many different purposes.

Thus, although Sprat's *History* (1667/1958) was promoted as a description of the Royal Society, and clearly expressed the aims and ideals of certain of its prominent members, it cannot be read simply as a statement of the commonly agreed norms of the organization. Paul B. Wood (1980) has argued that the text was an avowedly apologetic document, aimed at replying to criticisms of the Society from various quarters, and that tensions and even contradictions in the doctrines it contains can be ascribed to its origin in these circumstances. The book was therefore, as

Figure 1. Frontispiece from Thomas Sprat, *History of the Royal Society* (1667). The Royal Society has been the focus of a number of recent studies of the part played by institutions in the formation of the identity of the early-modern scientific practitioner. Reproduced by permission of the Syndics of Cambridge University Library.

Sprat himself put it, "not altogether in the way of *a plain History*, but sometimes of an *Apologie*" (1667/1958: B4v). Furthermore, as Michael Hunter (1989a) has pointed out, the work was not in any sense authored by the Society as a whole, being penned by an ambitious young cleric at the request of a small group of the fellows, and only rather distantly supervised by them in the course of its composition. Any attempt to read the text as an expression of programmatic norms would have to take account of the circumstances in which it was written; and similar strictures would apply to the interpretation of other declaredly apologetic works by such members of the Society as Joseph Glanvill, or indeed Robert Boyle.

Robert Iliffe (1992) has pursued the question of the authority within the Royal Society of the Mertonian norm of "communism," with specific reference to the dispute between Robert Hooke and Christiaan Huygens for priority as inventor of the balance-spring watch in 1675. For Merton, such a contest for ownership of intellectual property was a regrettable instance of "deviant behavior," the result of intolerable tensions between the norm of communicativeness and the countervailing valuation of individual originality. Iliffe suggests, however, that to refer to norms gives a misleading suggestion of the degree to which individuals were constrained by prevailing values:

> Where Merton writes of "norms," I refer in this paper to "conventions." Priority disputes depend upon the refinement and interpretation of sets of "rules" governing etiquette in publication – antagonists can appeal to these standards, they can reinterpret them, and they can create new ones. . . . Here I suggest that we should examine historically the ways in which actors manipulated these priority conventions in the process of achieving authorship of a text or ownership of an object. (Iliffe 1992: 30)

For Hooke, Iliffe argues, a set of conventions quite different from those commonly appealed to by natural philosophers was available, namely those prevailing in the mathematical and mechanical arts. According to these conventions, private ownership of intellectual property was quite acceptable and communication of an invention might legitimately be restricted, for example by writing in cipher. The more general point is that "Neither Hooke nor Huygens were bound by these conventions or codes, . . . rather they used them as it suited" (1992: 55). Iliffe's account establishes in detail just how many and how flexible were the resources that the disputants could call upon. They manipulated material objects, redrafted descriptions of them, respecified what had been said or thought or seen on particular occasions, and frequently revised their stipulations of what constituted the essential "idea" behind the invention and what might suffice as a "clue" to enable it to be copied. The outcomes of the dispute – the objects themselves, their verbal descriptions, and the assignments of credit for the invention – were all constructed through manipulation of these highly pliable conventional resources.

The behavior of Robert Hooke has been the object of a number of other recent studies, which have reconstructed aspects of his strenuous experimental work and his continuous negotiations for recognition and status within the Royal Society. While Hooke was the individual (along with Boyle) whose experimental interests contributed most to the research program of the Society in its early years, he was also a socially disadvantaged member of the group (Shapin 1989; Pumfrey 1991). Having no aristocratic lineage, and only dubious claims to gentility, Hooke was educated as a poor scholar at Oxford and lived thereafter by his wits as an inventor and architect. He worked with Boyle in the capacity of a servant on the Oxford air-pump experiments in the late 1650s and early 1660s and was relinquished by his patron to become the paid "curator of experiments" of the Royal Society in 1662. Although his skills earned him good standing as an experimenter, he continued to be treated by the governing elite of the Society in what he found to be an irritatingly high-handed manner, being required to produce experiments on a regular basis for their entertainment at meetings and occasionally admonished for failing in his duties. In his diary, he revealed a persistent anxiety that he was not being given appropriate respect by the other fellows; and he was never accorded the designation "philosopher" by them. Hooke's "role" in the Society was not granted automatically by virtue of his formal position there; on the contrary, he had to work very hard to gain the status and recognition he achieved.

Hooke, it is clear, did not find that membership in the Royal Society supplied him with a stock scientific role. Rather, he had to work with the material and cultural assets his context provided to construct such a role for himself. What was at stake, in effect, was not so much a role as an *identity*. Hooke's very personhood (including, as far as we can know them, his intimate anxieties) was shaped by his specific circumstances; but he was not just the passive object of forces impinging from outside. He was actively involved in constructing an identity for himself within the prevailing context of resources and constraints. He was, of course, limited by the range of conventional resources available to him, but he selected and reconfigured those resources in a creative and energetic way.

A picture of Hooke is emerging from these recent studies that is in line with a generally shared focus on individual motives and the construction of personal identities. Group norms are no longer seen to be transparently represented in such documents as Sprat's *History* and are found to have been ineffective in resolving at least some disputes. Michael Hunter, currently the most active historian of the Royal Society, whose work is not specifically constructivist in orientation, has confirmed this general direction by insisting upon the particularity and contingency of many of the developments of the organization's early years. Hunter has traced the perilous financial fortunes of the institution, the failures of its attempts to organize experimental research on a continuous basis, and the arguments among its members as to the direction it should take (1981: chap. 2; 1989c: 1–41). He cautions strongly against any assumptions that the institutionalization of experimental science was a straightforward process with a clearly defined and agreed-upon goal. No doubt these strictures against the teleology implicit in Ben-David's model are well taken, but, if pushed too far, they could leave the impression that the Society's survival and continuity were purely a matter of chance. A constructivist perspective would indeed seem to preclude any assumption that mature institutional forms or pre-given roles are inherent in the process of institutionalization, but historical inquiry can nonetheless take as its topic the creation of consensus and the cultural mechanisms that sustain an institution over time.

If we are looking for how communal values were constructed and sustained in the early Royal Society, we might consider the part played by Henry Oldenburg, whose career indicates how a sense of common purpose could be created through individual action. Oldenburg was a German immigrant who became a very active secretary of the Society in 1663 and held the office until his death in 1677. As well as maintaining a huge volume of correspondence with savants throughout Europe, he initiated and edited the Society's publication, *The Philosophical Transactions* (Oldenburg 1965–1986; Hunter 1989b; Shapin 1987). For many correspondents and contributors Oldenburg represented the corporate aims and values of the Society. He articulated the general notions of proper scientific method common among the leading members, keeping in check what the English saw as an unfortunate tendency to speculation and dogmatism on the part of some continental authors. He mediated disputes over priority or matters of fact, such as that between Johannes Hevelius and Adrien Auzout about the path of a comet observed in 1665. He instructed travelers to foreign lands as to what they should observe and how they should report it. Oldenburg clearly exerted considerable initiative in securing a crucial position for himself in the Society, but he did so by becoming the voice of the organization as a whole, and especially by representing it abroad. His secretarial labors played a significant role in creating an identity for the institution as a whole, and his editorship of the *Philosophical Transactions* helped to consolidate the rhetorical forms of scientific writing that long outlived him.

Research on the early Royal Society has formed only a part of a much larger enterprise devoted to reassessing the historiography of early-modern European science. In exploring the roots of the scientific movement in Renaissance and Baroque culture, constructivist perspectives have again been influential, but research that is not explicitly constructivist – which draws, for example, upon cultural history and anthropology – can be seen to be pointing in the same general direction. This research has considerably enriched our understanding of early-modern natural philosophy. It is beginning to unearth a complex archaeology of institutional settings within which the identity of the experimental philosopher was forged (cf. Porter and Teich 1992; Moran 1991a).

From this research, a much more differentiated patchwork of the social world of the early-modern scientific practitioner is emerging. The Royal Society was unquestionably an important site where legitimacy was conferred on the production of experimental knowledge, but it was by no means the only one and should not be regarded as the telos of a very variegated movement. A broad landscape of societies and academies has been mapped: some formally constituted, others informal and temporary; some national, others local; some dependent on the caprice of an individual patron, others sustained by a dedicated group of activists (Lux 1991; McClellan 1985: 41–66). Other, more traditional institutions have been shown to have been more supportive of scientific activity than had previously been recognized. Members of the Jesuit order were surprisingly active in natural philosophy throughout the seventeenth and eighteenth centuries (Dear 1987; Harris 1989). The medical profession was also heavily represented among experimental practitioners during the same period (Cook 1990). And universities, downplayed in the traditional historiography of the Scientific Revolution, were not at all barren fields for the growth of scientific knowledge (Brockliss 1987; Feingold 1984; Frank 1973; Gascoigne 1990). Finally, museums and private "cabinets of curiosities" are gaining greater attention as settings for the development of a novel empirical sensitivity to the natural world (Daston 1988; Findlen 1989, 1994).

A community dispersed over so many different institutional sites evidently did not share any ready-made identity. A "scientist" as such was not a recognized entity; so becoming one was not an option that was available to individuals working in these diverse settings. Instead, we find them patching together identities with the cultural resources that each location provided – as academics, physicians, gentlemen virtuosi, or courtiers, for example. Scientific practitioners in the early-modern period engaged in a process of "self-fashioning" at the same time that they worked to produce natural knowledge (Greenblatt 1980). Neither their "role" nor the knowledge they made was simply given by their environment.

They did not, however, work in isolation from others. Although the institutional structures of the modern scientific community should not be projected back upon this period, integrative forces that linked natural philosophers to one another have been identified by recent research. The correspondence networks initiated by the Renaissance humanists were continued into the seventeenth century by such copious letter-writers as Oldenburg, Marin Mersenne, and Samuel Hartlib. These overlapped with networks for gift exchange, whereby books, artifacts, and natural curiosities were passed from one person to another, generally ascending the social scale in return for favors granted (Biagioli 1993: 36–59; Findlen

1991). The operation of these gift-exchange networks was governed by the assumption that a rarity presented to a social superior would gain a reciprocal benefit for the donor. As late as the 1670s, samples of rare and striking phosphorescent substances were being presented to the English king, Charles II, who was said to "deserve . . . by understanding them, the greatest curiosities." Many gifts of "noble rarities" were sent to the Royal Society; some were repaid by election of the donor to a fellowship (Golinski 1989: 27–30).

Undergirding the exchange of gifts were structures of *patronage*. This form of relationship was pervasive in early-modern society and fundamental to the careers of most of those engaged in the making of natural knowledge. Mario Biagioli has written that patronage was "an institution without walls" in this period – a form of social interaction that was highly ritualized but not precisely localized. A patron conferred prestige and material support upon the client in return for gifts, amusement, enlightenment, or some other form of status enhancement. On the other side of the relationship, practitioners of the sciences engaged in identity formation or self-fashioning by intricate negotiations with their potential and actual patrons (Biagioli 1993: 1–30). Patronage remained a fundamental structuring factor in the scientific community throughout the eighteenth century. For most of that period the Royal Society was dominated by a succession of three great magnates: Sir Isaac Newton, Sir Hans Sloane, and Sir Joseph Banks. But in Renaissance and Baroque Europe it was royal, aristocratic, and ecclesiastical patronage that constituted the channels through which social power flowed. It was these forces that scientific practitioners sought to manipulate, and in relation to which they had to orient and represent themselves (cf. Eamon 1991; Moran 1991a; P. H. Smith 1991, 1994a, 1994b; Westman 1990).

The operation of the forces of patronage can be clearly seen in instances in which experimental practitioners worked in the setting of a prince's court. For example, Biagioli has analyzed the Accademia del Cimento, supported by Leopold de' Medici in Florence from 1657 to 1667, which yields a useful comparison with the early Royal Society. Being dependent upon the whims of its patron, the activities of this academy were more temporary and occasional than those of its more formally constituted London counterpart. Many of the experimental demonstrations discernibly partook of the character of courtly entertainments, diversions, or conversation pieces. The participants remained anonymous in the written account of the proceedings, an indication of their status as dependents upon the prince, by whom the legitimacy of the experimental claims was ultimately warranted. The contrast with the Royal Society is clear. For the London group, the king was a purely formal patron who attended no meetings. Instead, the independent status of gentlemanly experimenters was a crucial resource for certifying the validity of the facts they witnessed (Biagioli 1992: 25–39; cf. Findlen 1993a; Tribby 1991).

The most sustained discussion of the practice of natural philosophy in a context of early-modern patronage relations is Biagioli's *Galileo, Courtier* (1993). In this innovative work, the author draws upon constructivist perspectives along with other currents in cultural history and anthropology. He aims to show how Galileo's scientific practice and his self-fashioning as a natural philosopher and mathematician can be understood by reference to the setting of the Baroque court. The narrative follows Galileo's career at the court of Grand Duke Cosimo II of Tuscany, from 1610, and his subsequent involvement in maneuvers at the papal court of Urban VIII in Rome. Biagioli explains that his concern is with Galileo's construction of his own identity rather than with his biography as traditionally understood: "In a sense, Galileo reinvented himself around 1610 by becoming the grand duke's philosopher and mathematician. Although in doing so he borrowed from and renegotiated existing social roles and cultural codes, the socioprofessional identity he constructed for himself was definitely original. Galileo was a *bricoleur*" (Biagioli 1993: 3).

The term "bricolage," which Biagioli has adopted from anthropology and which he elsewhere defines as "an often opportunistic rearrangement of elements of pre-existing social scenarios" (1992: 18), is of key importance in his analysis. It suggests that Galileo did not invent his social world de novo; nor was he entirely passive in the face of its influences. He creatively combined and manipulated the elements his culture offered to fashion an identity for himself, claiming at court a standing as a mathematician that he could never have commanded in the Italian universities. At the same time, he produced natural knowledge, which in turn aided him in this process of self-fashioning.

A revealing instance is the crucial move to Cosimo's court in 1610, made possible by Galileo's discovery of the moons of Jupiter early that year. Obviously, Galileo did not imagine these objects into existence – he really saw them through his telescope – but he did nonetheless construct them as discoveries, representing them as appropriate "gifts" for the grand duke and simultaneously presenting himself as worthy of patronage as their discoverer. Biagioli shows how Galileo packaged his discoveries to maximum advantage. He named the moons "Medicean stars," proclaiming that these attendants of Jupiter (which was said to be astrologically associated with Cosimo) showed that the heavens themselves displayed the glory of the Medici. So successfully were the satellites linked to Medici prestige that they became incorporated in medals, frescoes, and even court masques (Biagioli 1993: 103–157).

Having traded his dazzling gift for Medici patronage, Galileo found himself expected to engage in disputes at the Florentine court. He became embroiled in a succession of debates on buoyancy with Aristotelian natural philosophers from the faculty of the University of Pisa. Biagioli argues that these disputes should be interpreted as rhetorical duels of honor, the point-scoring parries being enjoyed as entertainment by the patrons of those who took part. Patrons, however, could not intervene decisively without risking drastic loss of status if they were shown to be mistaken. Courtly disputes were therefore typically inconclusive and unproductive of any consensus as to the facts. Biagioli shows that this was true also of Galileo's debate about comets with the Jesuit mathematician Orazio Grassi, in which his striking literary skills and sideswipes against pedantry earned him the admiration of the pope (1993: 159–209, 267–311).

It was in the context of such disputes that Galileo gradually developed his position on the physical truth of the Copernican cosmology. Biagioli argues that Copernicanism was not a preexisting commitment for Galileo; it was not a fundamental philosophy from which all of his behavior can be derived. Rather, he came to articulate his Copernicanism as he strove to legitimate his status as a mathematician vis-à-vis the natural philosophers. In this context, arguing that the heliocentric cosmology was physically true helped to establish that mathematicians could be competent philosophers, and hence could escape the subordination to which the hierarchy of academic disciplines would confine them. Biagioli's claim here does not possess the clarity of a straightforward causal argument; he is not suggesting that the court context forced Galileo to adopt the Copernican theory. Nonetheless, Galileo's having taken up Copernicanism is said to have helped raise his status at court. Biagioli writes, "Instead of looking for one *cause*, we may investigate the *process* of mutual reinforcement between Galileo's new socioprofessional identity and his commitment to Copernicanism" (1993: 226).

Biagioli brings his study to a close by examining Galileo's trial by the Roman Inquisition in 1633. He argues that the celebrated event can be reinterpreted when set against the background of patronage and court culture. The pattern he discerns is the "fall of the favorite" (1993: 313–352). Particularly in the nondynastic papal court, reversals of fortune could be rapid and dramatic. A favorite client could be an asset to the patron for a while, but might have to be sacrificed to maintain authority in a crisis. This pattern does not, however, fit Galileo precisely. Although his relationship with Urban VIII had certain features of the type, Galileo was never an exclusive favorite of the pope. Nor can the complex and multiple issues raised at his trial be reduced to a matter of personal betrayal. As Biagioli recognizes, the trial has still to be interpreted in relation to other contexts – theological, philosophical, and diplomatic.

The trial nonetheless serves as an appropriate coda to Biagioli's book, its outcome – Galileo's conviction and forced recantation of the Copernican theory – having negated his painstaking self-fashioning as a courtier over the preceding few decades. The event also highlights some of the subtler nuances of the concept of self-fashioning. For Biagioli, Galileo is not to be seen as a fully autonomous subject who utilized patronage as a resource to pursue his already existing aims; his choices and even his motives were shaped by his context. And, as the episode of the trial spectacularly showed, Galileo also experienced the constraints of that identity in ways that eventually thwarted his desires. Biagioli's analysis asks the reader to accept a model of identity formation in which individuals are shaped in their intimate desires by their context and yet are capable of manipulating relationships (even those in which they hold a position of inferior power) in order to pursue their aims. Utilizing the resources of their culture to create their identities, they are nonetheless liable to experience changed circumstances as constraining their actions. Such a model seems to invoke a range of aims and motives, some long-lasting and some temporary, some self-generated and some formed by circumstances. It also implies ambiguous and variable relationships between individual and context, such that the setting may yield resources to be exploited or set up obstacles to be negotiated. A model as complex as this could undoubtedly benefit from more precise articulation and further application to historical case studies.

Difficult though they are, these questions of how personal identity is formed appear to arise unavoidably once we call into question the Mertonian assumption that the social role of the scientist unfolded unproblematically and teleologically. If we commit ourselves to investigating how the early-modern practitioners of the sciences constructed their identities, then we have to deal with complex interactions between modes of self-presentation and sociocultural settings that were themselves in flux. Problems concerning the range of resources individuals might draw upon and the constraints they might encounter demand some degree of theoretical articulation; but they can only be approached in the context of empirical research into particular cases. Rather than proposing to solve these problems here, then, I shall conclude this section by pointing toward two kinds of resources that may help with further work on the topic.

First, there are the intimate written records already used by conventional biography. Many early-modern intellectuals have left texts in which they recorded their day-by-day actions and thoughts or deliberated on questions of moral identity. The Christian traditions of confession and examination of conscience, and humanist rhetorical forms such as the essay and the commonplace book, were frequently drawn upon in composing these documents. Hooke's diaries and Boyle's moral essays have already been subjected to scrutiny from a biographical point of view.

As Shapin has pointed out, however, exploration of the construction of a social identity, though it may draw upon the archival materials used by biography, does not entirely share its aims. For one thing, a continuity of themes through the course of an individual life should not be assumed. It may be that circumstantial contingencies or temporary and local constraints are more consequential for how identity is expressed than childhood experiences or early ambitions. In addition, the status of the documents used must be considered in relation to their actual or implied audience and the conventions that shaped their form and content. Rather than being looked *through* as a window into the soul of the writer, such documents might more pertinently be viewed as exercises in which personal identity was molded by a process of working through putative modes of public self-representation. Read in this way, the essays written by Boyle in the late 1640s enable one to trace his painstaking creation of his own identity by developing means to present himself as sincere and disinterested, and by measuring himself against the available social roles of "gentleman," "philosopher," and "Christian virtuoso" (Boyle 1991; Shapin 1993, 1994: chap. 4; cf. Shapin 1991).

Second, studies of the issue of gender in early-modern science have also produced valuable resources for thinking about identity formation. Pioneering works in this field, by scholars including Carolyn Merchant (1980), Evelyn Fox Keller (1985), Londa Schiebinger (1989, 1993), and Ludmilla Jordanova (1989) have succeeded in establishing the importance of gendered imagery in the discourse of the sciences in early-modern Europe; they also suggest that gender might be a significant category for understanding masculine self-fashioning. The fact that almost all the practitioners of the sciences were men, long regarded as somehow "natural" or unimportant, has been shown to have been highly consequential. The careers of those few women who did succeed in penetrating the male preserve of scientific institutions can be very revealing of the means by which male identities were constructed in those settings. Examples of women whose careers have been scrutinized in this light include the seventeenth-century English natural philosopher, Margaret Cavendish, the eighteenth-century Italian physics professor, Laura Bassi, and the eighteenth-century French mathematician, Emilie du Châtelet (Schiebinger 1989: 47–65; Findlen 1993b; Ehrman 1986; Terrall 1995). These were among the women who achieved some recognition in the scientific communities of their time and carved out roles for themselves (albeit circumscribed ones) in their institutions. Studies of them have illuminated their strategies for success, the family and personal connections upon which they were able to draw, and the masculine prejudices they encountered and circumvented.

But the value of such studies extends beyond the specific cases of these individuals. Their importance is to shed further light on the process of self-fashioning of scientific practitioners by showing how centrally gender was embedded in the process. Modes of self-presentation within the scientific community were intimately tied to models of masculinity, whether the connection was made via notions of "honor," "indepen-dence," or "civility." Different constructions of the identity of the natural philosopher drew upon different notions of what it meant to be a *man*, from the swaggering deportment common in the Italian courts to Boyle's displayed modesty and celibacy (Tribby 1992; Haraway 1996). They also exploited gendered language, such as that pioneered by Francis Bacon, which described inquiry into the natural world in terms of penetration and coercion. Images of nature as a female, unveiled by male researchers, had a pervasive ideological role (Schiebinger 1989; Jordanova 1989). The counterpoint to this masculinist discourse was the articulation of theories of female physique and psychology, which had as one of their functions the provision of explanations of why women were naturally unsuited to contribute to the sciences.

It seems clear that gendered discourse and behavior entered into the self-fashioning of male practitioners of the sciences in a quite intimate way. In exploring this further, constructivist research can benefit from more contact with gender studies. Hitherto, relations between the two fields have not been consistently close; but the focus on identity forma-tion offers a shared domain of interest. In recent years, constructivist sociology of science has abandoned the notion of a pre-given social role toward which historical development has tended and which institutions have conferred upon people unproblematically. Instead, scientific prac-titioners are thought of as self-fashioned individuals who creatively ma-nipulate the resources offered by their cultural settings to form their own personae. A parallel trend has characterized gender studies, which have broadened from an initial concentration on the history of women to ex-amine the pervasive importance of gender as a category of culture and identity. Rather than viewing women's identities as fixed by nature, re-cent studies have insisted on their character as historical constructions of particular cultures. This development has historiographical implica-tions for the study of men too, who can be shown to have identities no less shaped by features specific to their gender. Further work on this topic could offer new resources for understanding the formation of per-sonal identity, even in a field as predominantly masculine as early-modern science.

THE DISCIPLINARY MOLD

Whereas the traditional "big-picture" accounts posited science as a single entity with a continuous history, historiography since Kuhn has high-lighted discontinuities in its development, even if Kuhn's invitation to tell the story of successive revolutions has not been accepted. In particular, the breach that separates early-modern science from that which emerged in the early nineteenth century has become more apparent. The period from about 1780 to 1850 has been labeled the "second scientific revolution" (Kuhn 1977a: 147, 220; Hahn 1971: 275–276; Brush 1988). This was a time in which new scientific disciplines such as geology, biology, and physiology were founded and existing ones (especially physics and chemistry) dramatically reconfigured. Remarkable changes in conceptual content and practice occurred in institutional settings that were themselves being transformed. In France, scientific and medical training was radically reorganized in institutions founded or reformed during the revolutionary era. In Germany, the universities were fundamentally reshaped, with revised academic rituals and a novel setting for experimental training: the research institute. In Britain, the Enlightenment tradition of genteel individualism continued to flourish, but the formation of a wide range of new educational and research institutions also testified to the impact of rapid technological and social change.

William Whewell's coining of the term "scientist" in 1833 might be taken to be emblematic of these changes, and the word is indeed more applicable to scientific practitioners in that period than to those of the century before. But Whewell's ambition to contribute to many different fields, and to attain recognition as a philosopher and historian of the sciences in general, was not typical (Yeo 1993). Instead, the boundaries of distinct disciplines became a more entrenched feature of the production of knowledge, embodied in the constitution of university departments and institutes, in specialized scientific societies, and in new journals. This was not simply a matter of a progressively finer division of a stable map of knowledge, as might be suggested by use of the term "specialization." Rather, entirely new domains of knowledge (unrecognized, for example, in Enlightenment encyclopedias) were being carved out, with their own defining practices and regulated borders. Disciplinary boundaries, although they were sometimes to be breached or relocated, became a structuring feature of natural knowledge in all of the places in which it was made (Stichweh 1992).

The institutional dimension of these changes has sometimes been described as a process of "professionalization." The term suggests that science as an occupation achieved a certain independent status in this period. A specific set of developments is usually pointed to as signs of this: regular programs of training, secure sources of employment, control by scientists themselves of the quality of their work, and an ideal of service to society at large. The model of "professionalization" does, however, have limitations of the kind that we noted in connection with the Mertonian claim that science progressively achieves "autonomy" from external influences. It implies that societies progress along a single path, on which the formation of a profession from the members of a distinct occupational group is an important landmark. By suggesting a pattern of development of this kind, the term carries connotations of teleology – the implication that the change can be explained by its orientation toward a particular goal.

Acknowledging this drawback, J. B. Morrell (1990) has suggested that the notion of professionalization can still be useful, if employed with sensitivity, to isolate important changes in the social profile of science in this period. Morrell argues that one should not assume that the transformation occurred simultaneously, or even in quite the same way, in every country or every scientific field. Nor should it be assumed that achievement of professional status was the *aim* of any group. Nonetheless, application of the model would suggest at least six features of change that we might expect to find – and do in fact find – in Europe and the United States in the first half of the nineteenth century. They are: (1) increased numbers of paid posts for scientific specialists at public and private institutions; (2) the rise of specialist qualifications, such as the Ph.D. degree; (3) an expansion of programs of training for students in research laboratories; (4) increased specialization of publications; (5) the rise of institutions, such as the British Association for the Advancement of Science (from 1831), which were organized to defend the interests of those pursuing scientific careers; and (6) the creation of an autonomous reward system for career scientists within their own institutions. A further feature might be added to Morrell's list: the rigorous exclusion of women from participation in the scientific professions. It has been argued that demarcation of the domain of the male professional scientist set up a new barrier against women's pursuit of scientific careers. A firmer boundary between professional specialist and public "amateur," for example in botany, tended to restrict the possibilities for women's contributions, curtailing the opportunities that a few women had been able to seize in the previous century (Schiebinger 1989: 244–264; Shteir 1996: chap. 6).

There is little doubt that these developments occurred, but the question remains how helpful it is to categorize them as aspects of professionalization. However sensitively applied, the term still suggests a process that is essentially common to many different disciplines, whereby they achieve autonomous control over their own affairs. The implication is of a social process, but one that is apparently oriented toward a goal of independence from "external" social forces. The problem of neglecting the specific practices of the different sciences, which we noted as a feature of Mertonian sociology, is evident also here. The constructivist perspective suggests that an alternative scheme is needed, one that would be sensitive to local settings, including the specifics of experimental and pedagogical practices, but would explore connections between developments at different sites. In place of an implicit teleology, we might consider how change might be explained as the unintended outcome of the interaction of uncoordinated, even conflicting, forces.

One plausible category for a constructivist analysis is "disciplinarity." By this is meant not simply the reconfiguration of the scientific disciplines in this period but also the embeddedness of this process in larger formations of power. Discipline formation requires the consolidation of a community that shares a particular model of practice, which in turn implies modes of regulating behavior that may have had wider application, for example in schools, prisons, or factories. The ambiguity of the word "discipline" is crucial here: It refers both to a form of instruction to which one submits and to a means of controlling behavior. The word has this dual significance deeply impacted in its etymology. It is worth recalling this original connection with pedagogical power, lest we assume too readily that disciplinary divisions of knowledge are simply "natural" (Shumway and Messer-Davidow 1991; Hoskin 1993).

The recognition of separate scientific disciplines is at least as old as Aristotle. Historians have studied the different ways of classifying the disciplines, and the techniques used to define their practices and police their boundaries, in the premodern and early-modern periods (Livesey 1985; Westman 1980; Dear 1995b). Kuhn used the term "disciplinary matrix" to characterize the conceptual elements that together make up "the common possession of practitioners of a particular discipline" (1962/ 1970: 182–187; cf. Kuhn 1977b). The claim of those who consider the second scientific revolution in terms of disciplinarity is not that disciplines as such were new, or simply that new disciplines came into existence, but that techniques for inculcating and perpetuating disciplines – for disciplining their practitioners – were transformed. Consideration of the changes of the period in these terms offers the promise of connecting a wide range of empirical studies – of scientific education, research training, laboratories, instrumentation, disciplinary rhetoric, and so on – with potentially illuminating theoretical perspectives. Most prominent among the latter is the work of the philosopher and historian Michel Foucault, whose influence is reflected in much of the recent work on disciplines. As Jan Goldstein (1984) has argued, Foucault's idea of disciplinarity offers an alternative to the model of professionalization for understanding the social dimension of scientific expertise and its extension through society. Although Foucault's work as a whole is not readily assimilable to the constructivist perspective, his thinking about disciplines is worth attention for this reason.

Foucault introduced the topic in his inaugural lecture at the Collège de France in 1971, when he suggested that disciplines offer a preferable alternative to ways of organizing the history of thought by reference to ideas, traditions, or authors' works – all categories he believed to be contaminated by the assumption of the "sovereignty of the subject," namely, the notion that thought was the product of autonomous and fully self-conscious human minds (Foucault 1971/1976). Disciplines comprised one of the "systems for the control and delimitation of discourse" that regulated what could be said in a certain discursive setting. Foucault wrote: "For a discipline to exist, there must be the possibility of formulating – and of doing so ad infinitum – fresh propositions. But . . . a discipline is not the sum total of all the truths that may be uttered concerning something." Disciplines contain errors as well as truths; and propositions licensed by them must satisfy conditions other than truth or falsity. They must refer to a certain range of objects, make use of conceptual techniques of a well-defined type, and fit into a certain field of theoretical understanding. Outside the range of a discipline are statements that are not so much false as categorically unthinkable: "Within its own limits, every discipline recognizes true and false propositions, but it repulses a whole teratology of learning" (1971/1976: 220–223).

This understanding of disciplines, still framed in terms that were largely internal to intellectual history, was broadened in Foucault's subsequent *Discipline and Punish* (1978). There he explored connections between the constraints operating on discursive production and the workings of power in settings apparently removed from the academic – the factory, the hospital, the prison. His central focus was on the creation of a new regime of penal practice in the early nineteenth century and on the discourses of the "human sciences" articulated around it. He firmly rejected the suggestion that this radical shift in practices of punishment reflected a straightforward "humanization." The dramatic public executions, involving mutilation and torture, which were characteristic of the eighteenth century, were indeed barbaric by the standards of later times (as Foucault showed with his graphic description of the execution of the regicide Damiens in 1757); but they obeyed a logic of their own and served a definite political purpose in ancien régime society. And the institution of the prison that increasingly replaced them from the beginning of the nineteenth century was no less the product of a regime of power – power that was applied to regulate, examine, and constantly scrutinize the body of the criminal.

And not just the criminal. For, according to Foucault, the techniques of disciplinarity put into practice in penal reform had a much wider application. They were used to produce "docile bodies" in armies, factories, schools, and hospitals; and to produce knowledge of the persons thus regulated. The "micro-physics of power" was, at the same time, a means of knowledge production – a battery of techniques for generating standardized information about disciplined individuals. The techniques of control included means of subdividing space and time, the remodeling of bodily gestures, and the restructuring of institutions to function like machines. The practice of surveillance or "hierarchical observation" was built into the very architecture of certain institutions. The "panopticon," a design for a circular prison in which every prisoner would be observable from a single point, proposed by Jeremy Bentham in 1787, is thus "the diagram of a mechanism of power reduced to its ideal form" (Foucault 1978: 205). Foucault concludes: " 'Discipline' may be identified neither with an institution nor with an apparatus; it is a type of power, a modality for its exercise, comprising a whole set of instruments, techniques, procedures, levels of application, targets; it is a 'physics' or an 'anatomy' of power, a technology" (215).

Disciplines, then, are understood by Foucault as apparatuses of power that function to produce knowledge about the human world they bring under control. The material architecture of prisons, which allowed for the localization, regulation, and surveillance of prisoners, expressed the same relations of power as the human sciences in which knowledge of the criminal personality or the sociology of crime was produced. The technology of control was of a piece with the technology of knowledge production. And the knowledge produced included the self-knowledge of the persons subjected to this technology of power: their "subjectivity" was a product of their "subjection." As an example of the techniques involved, Foucault considers the procedure of the examination:

> [W]ho will write the . . . history of the "examination" – its rituals, its methods, its characters and their roles, its play of questions and answers, its systems of marking and classification? For in this slender technique are to be found a whole domain of knowledge, a whole type of power. One often speaks of the ideology that the human "sciences" bring with them, in either discreet or prolix manner. But does their very technology, this tiny operational schema that has become so widespread (from psychiatry to pedagogy, from the diagnosis of disease to the hiring of labour), this familiar method of the examination, implement, with a single mechanism, power relations that make it possible to extract and constitute knowledge? (Foucault 1978: 185)

Techniques of this kind were precise in their points of application but readily utilized in many different settings. They marked the extension of disciplinary forms well beyond the confines of the academic world. Many local configurations of power assumed a common pattern, with common techniques of knowledge production. The process is seen by Foucault as responding to broad historical causes, such as demographic growth and changing relations of production, and as drawing upon traditions of discipline in institutions as diverse as monasteries and the army. But he denies that it can be reduced to the diffusion of power from a single source, such as the state or the ruling class. Power, for Foucault, is not something that is concentrated in one place from which it diffuses to others; it is "strategic" and "capillary," constantly reenacted through the disposition of actors in specific local situations.

As disciplinary practices were instituted in the sciences, they subjected both the things that were to be known and the potential knowers to new forms of discipline. Joseph Rouse has pointed out that the techniques which permitted construction of knowledge about the disciplined human subject – such as isolation, documentation, examination, and what Foucault calls "normalization" (the judgment by reference to a standard that imposes an implicit valuation) – were also applied to nonhuman objects in the laboratory. "The world we inhabit is riven with enclosures, partitions, and purifications, marked by measurements, counts, and timings, and tracked by new forms of visibility, documentation, and accounting, all in order to make scientific knowledge possible" (Rouse, 1993a: 143). Practitioners themselves were disciplined by their involvement in this regime. Students being trained in the sciences were supervised in their physical deportment and gestures, required to report their results in standardized forms, and subjected to examinations. Foucault's emphasis on the physical basis of relations of power suggests that we examine the architectural settings in which new practices were taught and the shaping of students' activities by the requirement to manipulate apparatus in specified ways. All of these components of a disciplined research training in the experimental sciences have in fact been shown to have emerged in the early nineteenth century, initially in the reformed German universities and subsequently at many other institutions in Europe and the United States.

The origins of research training in the sciences remain, in some respects, unclear. But William Clark (1989) has traced one of its roots to the style of teaching pioneered in the seminars for classical philology founded in Germany in the eighteenth century, especially at Göttingen and Halle (cf. Turner 1971). Practices of examination were integral to these institutions on several levels: Students were examined by the directors, who were in turn obliged to report in a standardized form to the state bureaucracies. Students' work habits were disciplined to an unprecedented degree by the requirement for them to submit regular written work for criticism. A structure of "hierarchical observation" was thus instituted by the circulation of paper documents.

By the second decade of the nineteenth century, the spread of these educational practices through the German university system had issued in the formal recognition of the Ph.D. degree, awarded for a substantial body of written work that was the result of original research. The requirement of originality (a novelty in the academic world) and the stipulations for the written form of the dissertation reflected, according to Clark, "a simultaneous aestheticization and bureaucratization of the degree candidate" (1992: 99). This demand that a candidate exhibit some depth of subjectivity by producing original ideas was first made in the context of the philological seminar. It offers itself to an analysis in Foucault's terms, according to which disciplinary structures are largely responsible for inducing the cultivation of a subjective sense of the self. As Clark puts it:

> Although cast into types by the routines and reports of the seminar as pedagogical institute, the seminarians were, nonetheless, condemned to a domain of autonomy. Ministry and directorate compelled the seminarian to acquire an original personality. . . . [He] fashioned himself as a routinely normalized yet peculiarly differentiated individual. . . . He had to articulate a sphere of private academic interest, and must transfix this persona for evaluation in writings. (Clark 1989: 127)

The scientific training subsequently put into place, both in university institutes and in seminars devoted to the natural sciences, subsumed the practices of hierarchical observation and examination. To the circulation of documents and the physical arrangement of pedagogical space was added a further element of the disciplinary complex – instrumentation. By learning to use apparatus, to produce the required experimental phenomena and to make observations and measurements, students learned routines of manipulation and acquired embodied skills. A portion of the didactic task was thus delegated to the material apparatus with which the students interacted.

A highly influential model of practical research training in the experimental sciences was provided by the institute for teaching pharmacists and chemists opened by Justus von Liebig at the University of Giessen in 1826. Liebig's institute has been taken by many historians as a paradigm "research school," particularly in the wake of J. B. Morrell's pioneering study (1972; cf. Geison 1981, 1993; Geison and Holmes 1993; Holmes 1989). It fulfilled the pattern described by Geison: "small groups of mature scientists pursuing a reasonably coherent programme of research side-by-side with advanced students in the same institutional context" (1981: 23). But, as Joseph Fruton (1988) has established, the majority of Liebig's students never became academic researchers. Those who achieved senior academic posts were greatly outnumbered by those who became pharmacists, and many others followed careers as physicians, industrial chemists, and high-school teachers. Fruton concludes, "the main educational function of the Giessen laboratory throughout the period 1830–1850 was the training of future pharmacists and industrial chemists" (1988: 17). For these students, any experience of participation in cooperative research was partial and temporary.

Nonetheless, for all of his students, Liebig's mastery of the practices of disciplinary training was remarkably effective. These practices included rigorous division of the students' time, weekly examinations, and a schedule of set tasks for them to perform, using the methods of organic chemical analysis that Liebig had significantly refined (Holmes 1989:

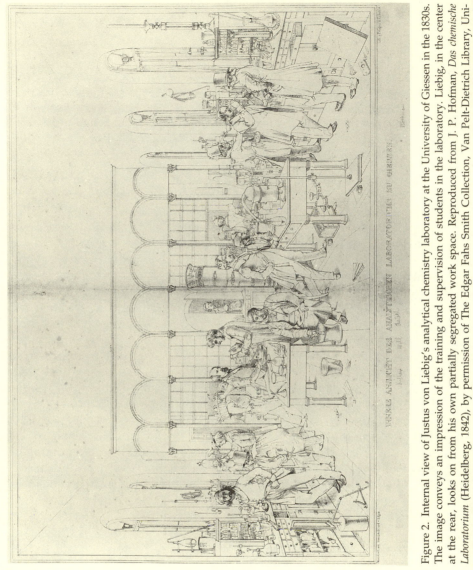

Figure 2. Internal view of Justus von Liebig's analytical chemistry laboratory at the University of Giessen in the 1830s. The image conveys an impression of the training and supervision of students in the laboratory. Liebig, in the center at the rear, looks on from his own partially segregated work space. Reproduced from J. P. Hofman, *Das chemische Laboratorium* (Heidelberg, 1842), by permission of The Edgar Fahs Smith Collection, Van Pelt-Dietrich Library, University of Pennsylvania.

127–128). In this well-defined field of experimental work, students were delegated tasks suitable to their level of skill; after 1832, the most advanced of them could publish the results of their researches in Liebig's own journal, the *Annalen der Chemie* (Morrell 1972: 30). Liebig also worked on modifying the physical space of his laboratory to facilitate the manual work of the students while maintaining the necessary supervision. A drawing of the laboratory as it was remodeled in 1839 shows students working in fairly close proximity – some apparently in pairs – at benches fitted up for the task. Liebig himself looks on from his own workspace, slightly separated and partially screened off from that of the students (Holmes 1989: 159; cf. Geison and Holmes 1993: frontispiece). Liebig later recalled his panoptic surveillance of his pupils' work with an appropriate metaphor: "I gave the assignments and supervised the execution; like the radii of a circle everybody had a common center" (quoted in Fruton 1988: 18).

Also of key importance in the disciplinary complex of Liebig's laboratory was an instrument: the combustion apparatus for quantitative analysis of organic compounds, which Liebig designed as an improvement on the previous model of J. L. Gay-Lussac and J. J. Berzelius (Holmes 1989: 131–142). As Morrell notes, "With practice, any determined student could obtain reliable results with Liebig's combustion apparatus" (1972: 27). By learning, under supervision, to master the apparatus, the student gained command of formal method and tacit skills; he assimilated at the level of bodily gesture an account of the chemical basis of the composition of organic substances and a procedure for determining it quantitatively. Part of Liebig's pedagogical mission was in effect discharged by his hardware – the instrument itself took on some of the task of disciplining its users.

As the disciplinary model of research training was adapted elsewhere, other ensembles of apparatus assumed this function. The physiological institutes founded at many German universities around the middle of the nineteenth century relied upon a range of instrumental systems applied to animal and plant specimens, from microscopic examination to more elaborate preparations of living tissues in which electrical or chemical stimuli were applied and their effects measured. The science of physiology essentially depended upon this kind of experimental arrangement, in which phenomena and instruments were packaged together (Lenoir 1986; Coleman and Holmes 1988b). Students were introduced to the subject by learning to manipulate these experimental setups. The first physiological institute in the German-speaking world – that established by Jan Purkyne in Breslau in 1839 – initiated the tradition of hands-on training in the use of apparatus. In Purkyne's case, the approach was inspired by the emphasis on learning-by-doing and bodily exercise of the Swiss educationalist Johann Pestalozzi (Coleman 1988). When microscopes of improved design became available in the 1840s, they assumed a central role in Purkyne's laboratory training and in that of such colleagues as Jacob Henle at Heidelberg (Tuchman 1988). In the following decade, William Sharpey, at University College London, had a microscope mounted on a rotating stand so that it could be passed from student to student at the laboratory bench (Geison 1978: 55). And by the 1870s, Thomas Henry Huxley had incorporated microscopes into structured disciplinary training at the Royal School of Mines. Huxley explained how he drilled his students to observe with the microscopes in the correct manner:

> We have tables properly arranged in regard to light, microscopes and dissecting instruments, and we work through the structure of a certain number of animals and plants. . . . [T]he student has before him, first, a picture of the structure he ought to see; secondly the structure itself worked out; and if with these aids, and such needful explanations and practical hints as a demonstrator can supply, he cannot make out the facts for himself in the material supplied to him, he had better take to some other pursuit than that of biological science. (quoted in Gooday 1991: 339–340)

In subjects other than physiology the instrumental component of the disciplinary complex may have been less important. Kathryn Olesko (1991) has analyzed, in great detail, the physics seminar of Franz Neumann at the University of Königsberg from 1834 to 1876. Here, cultivation of personal originality took second place to fairly rigid inculcation of manual skills. The aim was not so much *Bildung* (humanistic self-realization) as *Ausbildung* (training or drill). Neumann pursued this pedagogical aim through imparting methods of precision measurement that were applicable to many different experimental phenomena. Olesko focuses particularly on his teaching of procedures of error analysis, using such techniques as the method of least squares. For Neumann, mastery of such techniques would give the prospective physicist or physics teacher fundamental skills that could be applied to a wide range of problems. Although they were taught in a setting of intensive experimental work, with apparatus specifically developed to measure phenomena in mechanics, optics, or electricity, the methods themselves proved remarkably mobile and were readily taken up elsewhere (Olesko 1991: 13–14, 44–47, 311–312; cf. Olesko 1993).

Olesko thus draws attention to the transmissibility of certain techniques that could be readily articulated. Neumann's methods were mobile because they could be spelled out in detail in written documents; they were detachable from any particular experimental apparatus, with its inevitable complement of tacit manual skills. The contrast with knowledge that was more directly tied to instrumentation or bodily skills, and hence more rooted in a specific locality, is clear. This does not imply, however, that abstract methods are not linked to specific disciplinary complexes. They might indeed have been specifiable in ways that made them available for ready transfer to other sites, but they were only learned in connection with particular exemplars of practice and were assimilated elsewhere in, presumably, differently configured practical settings. Thus, Graeme Gooday (1990) has described the adoption of methods of precision measurement in physics teaching laboratories in Britain in the second half of the nineteenth century. In William Thomson's Glasgow laboratory, for example, as in Königsberg, precision measurement was valued as an essential component of a disciplinary training in physics. But, in Glasgow, measurement was tied directly to apparatus and problems connected with the Atlantic telegraph project, in which Thomson was significantly involved. In this context, methods assimilated from elsewhere were put to locally specific uses.

This is a pattern we might expect to find repeated. The fabric of a disciplinary complex – divisions of time and space, supervision of bodily deportment and manual skills, observation and examination, and specific instrumental ensembles – is tied to a particular locality. But recorded information and material artifacts can travel, and they may carry elements of the complex with them. They may even be put to work in settings well outside the laboratory. Thus, Gooday (1991) has traced the passage of microscopes into the schools and middle-class households of Victorian Britain. He shows that use of the instrument depended upon general cultural assumptions about nature that had already been widely diffused, as well as upon the efforts of those who manufactured, retailed, and advertised the apparatus and instructed purchasers how to use it. The structure of discipline in the domestic parlor was inevitably less rigid than in Huxley's teaching laboratory, but it was not altogether absent. The instrument itself conveyed some part of the technique needed to use it, but it also required extension of the disciplinary network to support it. The transition of microscopes into Victorian households both relied upon and extended the structures of disciplinarity built up in the pedagogical institutions of the experimental sciences.

The key to the commonality of disciplinary forms at different locations – which Foucault notoriously failed to explain – is perhaps to be sought in the transmission of those elements of disciplinary structures that can be made to travel. Constructivist sociologists from Fleck to Latour have engaged with these issues on a theoretical level, as the previous chapter showed. There is also, of course, scope for much further historical research on the local particularities of disciplinary training in the experimental sciences. We could do with more information about how bodily skills, material apparatus, and formal instruction related to one another in different disciplines at different times and places. Such further studies will enable us to approach the issue of the extension of disciplinary structures in a more informed way.

Although this chapter has highlighted disciplinarity as a key feature of the second scientific revolution, there is also scope for applying the concepts of identity and self-fashioning that have been put to use in understanding early-modern science. Doing so might require some departure from Foucault's analytical categories, however. His outlook views individual identity as the product of disciplinary structures of power, which are seen as constitutive of subjectivity. Even originality can be read, through Foucault's eyes, as an outcome of discipline. But the historian of modern science can readily point to examples of individuals who have fashioned themselves outside existing disciplines, at the crossover points between disciplines, or in opposition to the prevailing currents within their discipline. The computing pioneer Alan Turing, the physicists (like Francis Crick) who became biologists in the 1940s, or the maverick geneticist Barbara McClintock spring to mind (Hodges 1983; Keller 1983). These individuals' fashioning of their identities cannot be explained as the result of hegemonic structures of disciplinary power. Rather, some recognition of the capacity of individuals for autonomous self-expression seems to be required. Notwithstanding the powerful sway of disciplinarity over the personal formation of scientific practitioners, these cases remind us that individuals can construct their own professional identities by creatively manipulating the resources they find around them. Even in the age of disciplines, then, an understanding of scientists' identities may require reference to such notions as virtuosity and bricolage.

3

The Place of Production

Salviati: The constant activity which you Venetians display in your famous arsenal suggests to the studious mind a large field for investigation. . . .
Sagredo: You are quite right. Indeed, I myself, being curious by nature, frequently visit this place for the mere pleasure of observing the work.
Galileo, *Dialogues Concerning Two New Sciences* (1954: 1)

THE WORKSHOP OF NATURE

Chapter 2 ended by returning to the issue of localization in relation to scientific knowledge. The notion of a disciplinary matrix implies that experimental training occurs in a setting constituted by specific local circumstances. These might include such features as the spatial arrangement of facilities and the procedures that regulate human conduct and deportment. The reproduction of experimental effects elsewhere seems to require some transfer of elements of this matrix, whether in the form of texts, material artifacts, or embodied skills.

As Chapter 1 already indicated, this notion of the local specificity of scientific knowledge has been a common theme in constructivist studies. Kuhn's work first directed attention to the community of practitioners who share a "paradigm" for the understanding of nature. In Kuhn's view, natural phenomena are interpreted in a context that comprises certain presuppositions, methodological stipulations, institutional bonds, and so on, articulated around a core of a certain model of practice. The sociologists of scientific knowledge developed the topic of localization in a more concrete direction, highlighting the laboratory as a specific place of work. Their supposition was that experimental phenomena are produced and interpreted with the concentrated resources of particular physical sites. In the spaces designated as laboratories, they claimed, human and material resources are assembled to fix and construe phenomena and to package them for reproduction elsewhere.

The constructivist perspective clearly opens up for exploration the topic of the historical constitution of the laboratory. Among others, the following problems suggest themselves. What are the distinguishing characteristics of the laboratory, and how did the institution arise? How are its boundaries controlled to allow resources to be gathered but to protect them once they are there? How is communication of the results of laboratory work to the outside world managed? How does the physical site constrain the behavior of human actors, or at least define roles for them to occupy? And, finally, what comparisons can be made between the laboratory and other sites where natural knowledge is produced, including the museum, the clinic, the lecture theater, and the apparently more diffuse sites of the fieldwork sciences, such as geology, natural history, and ecology?

In this chapter, we consider these and other questions, initially focusing specifically on the laboratory and then widening the view to embrace other places in which natural knowledge is made. I shall review the attempts that historians have made to answer these questions and assess the implications of their findings for the proposition that natural science is a species of local knowledge. I should reiterate at this point that the historians whose work I shall mention would not all consider themselves specifically indebted to constructivism. Histories of particular laboratories, observatories, and scientific societies have a long and respectable lineage in the institutional history of science quite independently of constructivist sociology. And many of the new perspectives that have been brought to bear on these topics derive from other fields – for example, anthropology, where the notion of local knowledge has had independent currency (Geertz 1983), or architectural history. It is, nonetheless, a supposition of what follows that constructivist ideas have provided a theoretical rationale for this interest and offer the prospect of connecting empirical local studies with more general themes concerning the constitution of scientific knowledge (cf. Agar 1994).

To assert that science is a species of local knowledge is, of course, a rather more radical claim than to say the same of, for example, Azande witchcraft beliefs. Robert Merton's quip about the Nuremberg laws being impotent to affect the validity of the Haber process has much inherent (or at least inherited) plausibility. It gives epigrammatic force to what might be called the "everywhere and nowhere" view about science – that its validity extends universally while its place of origin is nowhere in particular (that is to say, no specially privileged place). It is in order to deconstruct the first part of this claim that constructivists have been working outward from laboratory studies to bring under scrutiny the wider realm in which scientific knowledge is taken to hold good. Their work could be said to have been directed at showing how this proclaimed universality might be the result of human action rather than divine gift.

Investigation of the laboratory is one way of addressing the second part of the claim, namely, that "the social place of knowledge is nowhere" (Shapin 1990: 192). Very long-standing traditions concerning natural and spiritual knowledge in Western culture portray it as the result of essentially mental processes that occur in a realm supposedly apart from the physical. These traditions have been reinforced by the resort of some practitioners of philosophical or spiritual disciplines to places of solitude. Taking themselves off from human society, whether to deserts, mountaintops, or stove-heated rooms, saints and sages have asserted that they thereby achieve closer contact with the realm of truth. It is indeed quite plausible to assume that a degree of privacy is conducive to intellectual deliberation, and it may even be the case that some experimental investigations require or benefit from relative seclusion. But these findings do not amount to an endorsement of the view that knowledge can be produced apart from human society: To be deliberately antisocial is not the same as being *asocial*. Even to seek solitude is to behave in accordance with certain social conventions, and such behavior may indeed serve the purpose of displaying that one is adopting a certain social identity. To make manifest that one seeks and relishes isolation is one way of assuming the role of a dedicated searcher after truth. As Shapin notes, "[S]olitude was often an intensely public pose. . . . It made no sense without that public audience, and its meaning depended utterly on a publicly understood language and stock of evaluations" (1990: 195).

To illustrate this, we can briefly consider two individuals whose deliberate cultivation of solitude had significant consequences for the public reception of their scientific work: Isaac Newton and Charles Darwin. The privacy sought by these two great scientists has sometimes been thought to remove their work from the possibility of sociological analysis. But such a supposition is mistaken, as recent historical treatments of them have shown. Instead, their solitude can be seen as having been made possible and meaningful by certain available cultural traditions.

Newton lived almost continually in Trinity College, Cambridge, from his admission as an undergraduate in 1661 until he left to take up a civil service position in London in 1696. He was not, of course, completely alone. College servants and an amanuensis ministered to his needs; in the early years his tutor and other dons took an interest in his intellectual progress; he was involved for a while in university politics; and, as Lucasian Professor, he occasionally lectured (albeit often, according to legend, "to the walls"). But, in certain respects, Newton chose an isolated existence and represented himself to his contemporaries as having done so. His communications to the Royal Society in the 1670s on the phenomena of light and colors portrayed him as a solo experimenter from whose "darkened chamber" supposedly incontrovertible results emerged. In the 1680s Edmond Halley acted as strenuous midwife to the *Principia*, which was seen to be drawn forth with great tact from the same fastness of solitude.

Newton took particular care to maintain his privacy in relation to his alchemical experiments. In a small laboratory built in the garden attached to his college rooms, he taught himself the basic chemical operations and tended his furnaces. During the decades of the 1670s and 1680s he also pondered at length many alchemical texts, compiling thousands of pages of notes as he attempted to decode the allegorical language in which he was convinced the authors had concealed their doctrine. This doctrine concerned "a more subtile[,] secret & noble way of working" than the common manipulations of "vulgar chymistry" (quoted in Golinski 1988: 151). Knowledge of spiritual alchemy, which he believed had been passed down as a secret tradition from ancient times, ought to be kept apart from the public gaze, in Newton's view. For this reason, he carefully cultivated a degree of solitude that continues to thwart historians who wish they could learn more about his contacts with other alchemical investigators.

The demarcation of a realm of proper philosophical secrecy, veiled from scrutiny by the "vulgar," also underlay Newton's stipulations as to how his work on mathematical natural philosophy should be understood. He saw no role in the appraisal of that work for any audience extending beyond a small expert elite. In the controversy that followed publication of his optical experiments in 1672, Newton insisted that his findings were "demonstrative" of his conclusion that white light was composed of rays of different colors (Schaffer 1989). This being so, neither replication of his experiments nor multiplication of witnesses was necessary. By denying the pertinence of repeated experiments or group witnessing, Newton was in effect questioning the model of science as a public activity upheld by Boyle, Hooke, Oldenburg, and others among the founding members of the Royal Society. To quote Shapin again: "While Boyle's and Hooke's public was accustomed to weigh, consider, and modify empirical claims, the public Newton wrote for was instructed to assent to proper mathematical demonstration. . . . The public that guaranteed scientific objectivity for Boylean experimentalists became, in Newtonian practice, a continuing potential source of corruption and distortion" (Shapin 1990: 206; cf. Iliffe 1989: chap. 6). Newton apparently turned his isolation to use, representing it to his contemporaries as a condition of work on recondite mathematical and theological problems.

When Darwin took up residence, in 1842, at Down House in Kent, he advertised to his contemporaries that he was seeking isolation from the urban bustle he had previously experienced in London. His withdrawal to rural seclusion signaled a retreat from immediate involvement with the scientific community of the capital. He wrote that he was settling at the "extreme verge of [the] world," although the site was in fact less than two hours' journey from the center of the city (Desmond and Moore 1991: 305). Soon after moving in, he announced that "the publicity of the place is at present intolerable" and set about increasing its privacy (306). He had workers divert the road away from the front of the house, at significant expense, and mounted a mirror on the wall outside his study window so that he could see who was calling at the front door.

Adrian Desmond and James Moore have plausibly linked these precautions to Darwin's anxiety about publicly revealing his unorthodox ideas about the transmutation of species. In a social milieu troubled by respectable fears of radical materialism, avowal of belief in the mutability of species was, Darwin wrote, "like confessing a murder" (Desmond and Moore 1991: 314). Somewhat more speculatively, Desmond and Moore suggest that the same anxieties contributed, at least in part, to Darwin's long-standing and apparently incurable ill-health. Darwin's isolation can thus be seen both as culturally coded and as a response to perceived social pressures. Representing himself as seeking rural seclusion, he mobilized the ancient traditions that identified places of solitude with the revelation of truth, as well as more immediate cultural resources, such as the role model of the country clergyman. Retreat to Down helped Darwin both to keep the outside world at bay and to prepare for revelation of his heterodox ideas on his own terms.

It was not, however, in any sense an asocial situation that Darwin occupied. He remained very well integrated within the scientific community, receiving frequent visitors and occasionally traveling to London or to scientific meetings elsewhere. He was an enormously prolific correspondent, not only with scientific colleagues but with a whole army of informants and suppliers of specimens scattered throughout the world. James Secord (1985) has shown how dependent Darwin was on information supplied by thousands of pigeon fanciers, horticulturists, beekeepers, and sheep and cattle breeders, whose experience and lore he translated into pertinent data. *The Origin of Species* itself encapsulates the whole social world of Darwin's informants, from "Mr Barrow," who reports on dog breeding in Spain, to "Mr H. Newman, who has long attended to the habits of humble-bees" (Darwin 1859/1968: 93, 125).

For both Newton and Darwin, then, seeking solitude was a means of regulating human intrusions, but it was not an asocial act. Seclusion had obvious practical benefits, in terms of reducing interruptions or invasive scrutiny of stigmatized intellectual activity. It also made sense according to certain cultural repertoires of meaningful behavior. To represent oneself as withdrawing from society was understood as a means of drawing closer to the realm of abstract truth. And the places to which they withdrew were recognized as appropriate for such an act as self-isolation in the pursuit of knowledge. Newton's college rooms and Darwin's country house were made, not just of bricks and mortar, but of the social conventions that validated those sites as appropriate for the production of natural knowledge.

Since the seventeenth century, the laboratory has come to be recognized as the preeminent site for making knowledge in the experimental sciences. It straddles the realms of private seclusion and public display, and calls for means of managing the transitions between them. On the one hand, the laboratory is a place where valuable instruments and materials are sequestered, where skilled personnel seek to work undisturbed, and where intrusion by outsiders is unwelcome. The Oxford dictionary defines it as "a building set apart" for experimental operations; it has been said to occupy one pole of a "continuum of relative privacy" (James 1989: 1–2). On the other hand, what is produced there is declaredly "public knowledge"; it is supposed to be valid universally and available to all. Experimental knowledge must therefore make the transition from the seclusion of the laboratory to the realm of public discourse. And the privacy of the laboratory, however practically necessary, is something of an ideological embarrassment.

The ambiguous situation of the laboratory between private and public realms was evident already at the time of its origin in early-modern Europe. As Owen Hannaway has documented (1986), the model of a site dedicated to manipulative experimentation was largely rooted in the traditions of practical alchemy. The word "laboratory" continued throughout the seventeenth century to be applied exclusively to rooms fitted with furnaces for chemical operations (Shapin 1988b: 377). This meant that laboratories carried with them associations of seclusion and obscurity – they were frequently sited in basements, which were metaphorically suggestive of the secrecy that alchemists were generally thought to practice. At the beginning of the seventeenth century, the chemist and humanist pedagogue Andreas Libavius castigated the Danish astronomer Tycho Brahe for maintaining his basement laboratory as a place of concealment. Tycho himself announced that the experimental knowledge produced there would have a curtailed circulation; he would be prepared to share certain things with princes or learned men, he wrote, "provided that I am convinced of their goodwill and of the fact that they will keep these things secret. For it is not expedient or fitting that such things become common knowledge" (quoted in Hannaway 1986: 598; cf. Shackelford 1993). The stereotype of the chemical laboratory as a secret place was invoked repeatedly through the following two centuries. In the 1750s the philosopher and economist Adam Smith was still denouncing the obscure discourse of "those . . . who live about the furnace" (1795/1980: 47).

This could be a troublesome heritage for those whose declared intention was to make experimental knowledge public. A rhetoric of "openness" had been voiced by certain practitioners of the technical arts since the Renaissance and was particularly common among authors of the books on technology that flooded from the printing press (Eamon 1984, 1990, 1994; Long 1991; Eisenstein 1979: 543–566). In the works of prophets of the new science, such as Francis Bacon and Thomas Sprat, the commitment to making experimental knowledge public was articulated as a goal. The Royal Society justified its practices of witnessing experiments, and other activities like compiling "histories" of arts and trades, in terms of this ideal of openness. As witnesses to experiments, readers of the *Philosophical Transactions*, discussants at meetings, correspondents, respondents to questionnaires, and reporters of phenomena, the audience for the new science was assigned an active role in its production, which was thus made manifest as a public activity.

Because it voiced, and to some extent practised, this ideal of experimental science as a public activity, the Royal Society was vulnerable to criticism on two fronts. On the one hand, Thomas Hobbes denied to the Society the status of a public place. He had found himself barred from admission and hailed this as evidence that the production of experimental knowledge was not truly open to all (Shapin and Schaffer 1985: 112–115). On the other hand, the attempt to construct knowledge in public ran up against the interests of those who wanted to maintain technical "secrets." Henry Stubbe (1670: 89), no less vehement a critic of the Society than Hobbes, raised the specter that its activities were *too* public to protect the interests of craftsmen and tradesmen. Their methods were being inquired into and, Stubbe implied, they had reason to fear that valuable information would be revealed to potential competitors. He suggested that reproduction of knowledge in the public realm would pose a serious threat to intellectual property rights.

These problematic issues were focused specifically on the laboratory – a place that was traditionally (and perhaps necessarily) private but that also had to be shown to be public. No perfect resolution of this dilemma was possible, but measures were adopted that, in practice, helped to defend and validate the laboratory's ambiguous status. Verbal descriptions and visual images were circulated to represent the contents of the laboratory and the results of the work that went on there, thus displaying its openness to the public sphere as a condition of the validity of the knowledge made therein. Procedures were also formulated for regulating access to laboratories. Both Boyle and Hooke were accustomed to managing access to their places of work. Boyle, whose laboratories were located in his residences, designated the rooms as spaces for private withdrawal from social intercourse, exploiting his culture's recognition that pious individuals needed a place of spiritual retreat. He found it necessary to insist that he was not available to all callers, but he did nonetheless admit to his laboratory, even at the cost of interruptions to his work, callers who had recognized social standing or who arrived with the appropriate letters of introduction (Shapin 1988b). Hooke's comparatively lower status was made manifest by his lack of similar means of control over access to his place of work. Because he was employed by the Royal Society, his work was vulnerable to inspection by the fellows. If this happened relatively rarely, it was probably because Hooke's quarters at Gresham College were not considered suitable places for the entertainment of gentlemen. Instead, Hooke was summoned to the homes of the more genteel fellows, such as Boyle and Christopher Wren, or was obliged to remove his apparatus to the meeting rooms of the Society to show his experimental findings (Shapin 1988b).

Boyle and Hooke were both engaged in utilizing available social norms to validate certain locations as sites where natural knowledge could be produced. They manipulated their contemporaries' recognition of certain types of places – gentlemen's houses, college rooms, monks' cells, artisans' workshops – to help secure credit for what they claimed about nature. They used to advantage the links between physical place and social conventions, while also making use of the pliability of material spaces to serve their aims. The seventeenth-century laboratory was not simply a college room, nor a prayer cell, nor exactly a workshop, but it was created from these places by rearrangement of the material fabric of the space and by exploiting the culture that governed how people perceived these places and how they acted there. The theoretical point, extracted by Ophir and Shapin, is that "Physical divisions in space – and whatever resistances they offer to social transactions – are saturated with culture and they exist as symbols within our culture" (1991: 10). It is because arrangements of physical space shape human activities, and yet are malleable to serve human purposes, that they could be adapted to facilitate the acceptance of claims to natural knowledge.

Since the seventeenth century, awareness of the special character of the laboratory, and some understanding of what goes on there, has become more widely diffused. Sharon Traweek, an anthropologist who conducted fieldwork at the Stanford Linear Accelerator Center in California, found that entry was regulated less by the "quasi-military" appearance and slightly lax procedures of the security guards than by a more subtly communicated awareness that strangers were not welcome to enter at will. Those outsiders who were admitted on regularly organized tours "often behaved as though they had been granted a special dispensation to see the inner sanctum of science and its most learned priests" (Traweek 1988: 19–23). Anthropologists such as Traweek have entered laboratories with the aim of dispelling this widespread quasi-religious awe of the places where science is practised.

The increased deference toward scientific institutions in the modern world can be ascribed in part to the physical presence of whole buildings dedicated to science. Inquiry into the architecture of scientific institutions is, however, a relatively recent development among historians. Sophie Forgan (1986, 1989, 1994) has explored the architecture of scientific societies, university laboratories, and museums in nineteenth-century Britain. She emphasizes that architectural form can be viewed as a series of responses to practical needs and as a symbolic accomplishment. The latter function may be discharged not just by ornamentation and "style" but by choices of site, building materials, means of access, and internal arrangement of facilities. Thus, Forgan notes that the buildings erected for scientific societies in Victorian cities represented "claims to a territory, both physically and metaphorically, and in a concrete sense embodied that claim and clothed it with institutional respectability. Public recognition followed hard upon the demarcation of institutional territory" (1986: 91).

Comparable studies of the architecture of scientific institutions include Mari Williams's (1989) on the Imperial Russian astronomical observatory at Pulkowa (Pulkovo) and David Cahan's (1989) on the Physikalisch-Technische Reichsanstalt in Berlin. In both cases, analysis disclosed both functional and symbolic purposes served by the choices of site, ground plan, materials, and style. Each institution can be seen to have been designed to provide convenient accommodation for the activities pursued there as well as to communicate a message to the world outside. The building at Pulkovo, near Saint Petersburg, was designed in 1834 with a facade chosen because it "clearly indicated, through its special character, the scientific role of the building" (Williams 1989: 120). That role included very accurate positional astronomy applied to problems of navigation and to the project of surveying the imperial territory. At the Physikalisch-Technische Reichsanstalt, erected in the 1880s, the symbolic message concerned German leadership in methods of precision measurement and their application to engineering. The internal arrangement of facilities at the two sites manifested the same dual purpose. At Pulkovo, the layout of rooms reflected the kind of astronomy practised there – with data reduction performed in immediate proximity to the observing instruments – and displayed differences in recognized status among the personnel, a feature that Traweek (1988: 33) found still evident in the architecture of scientific institutions today.

By mapping the internal arrangement of laboratories, Forgan has also illuminated the spatial dimension of disciplinary formations in the nineteenth century. As we saw in Chapter 2, spatial relations constituted a crucial element of Foucault's model of "discipline," most evidently in the paradigm case of Bentham's panopticon. Pursuing the question of the connection between structures of power and configurations of space, Forgan has noted (1989: 424) how teaching laboratories built in British universities in the nineteenth century mimicked Liebig's style of floor plan (mentioned in Chapter 2). Benches were moved away from the walls so that the instructor could stroll between them to monitor the students' work; they were also frequently aligned in rows so that they could be surveyed from a single position or oriented toward a source of natural light for microscopy. Lecture theaters, on the other hand, were built with tiered rows of seats, frequently arranged in the semicircular form of an amphitheater. Such an arrangement suggests a reversal of the Foucaultian panoptical gaze, since it is upon the lecturer that the students' eyes are supposed to be focused. But Forgan also noted the case of the chemistry class at the Liverpool University College, where laboratory workbenches were arranged in a tiered amphitheater surrounding the lecturer's podium. This arrangement indicates a large degree of surveillance from the center, in "an exceedingly didactic, controlled and organized form of teaching" (1989: 426–428).

"Disciplinarity," we recall, implies regulation of relations between personnel and material apparatus within a specific setting. The physical space of the laboratory provides means for organizing the interactions among its human inhabitants as they engage in experimental work. Historians have begun to consider the different ways in which labor has been organized in the laboratory. In contemporary institutions, such as that surveyed by Traweek, clear distinctions of status are evident between ancillary staff, technicians, research scientists, administrators, and secretaries. The status distinctions are directly tied to different roles in the work of the laboratory, but they are also reinforced by subtler codes, such as styles of clothing or the custom that different groups socialize at different places in the building. Many distinctions are also reinforced by gender lines. Women are commonly assigned to what are regarded as less prestigious occupations. More of them work as secretaries than as administrators; more as technicians than as scientists. The historical origins of these practices are gradually being unearthed.

Women do not appear to have been admitted to seventeenth-century laboratories, though they were occasionally present among the audience for scientific experiments. Boyle showed aesthetically pleasing phenomena involving phosphorescence and color changes in chemical solutions to female audiences, but this was a domestic display clearly differentiated from serious experimental work (Golinski 1989: 26). Technicians, on the other hand, were essential, albeit largely "invisible" in published accounts of experimentation (Shapin 1994: chap. 8). "Laborants" or "operators" might be men hired for their specific technical skills as chemists, glassmakers, clockmakers, or whatever, or they might be recruited from domestic servants. Lacking the social status of independent gentlemen, they were not generally credited with the capacity for original thought; nor were they assigned formal authorship, even in cases when they actually penned the written accounts of experiments. Reason (the work of

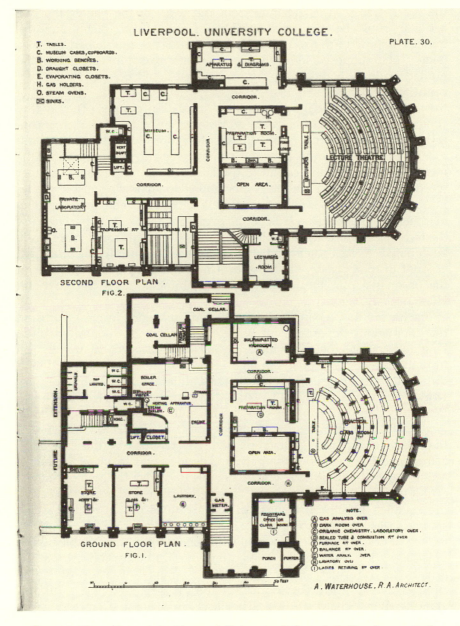

Figure 3. Floor plans of the Liverpool University College, including the practical chemistry class. Forgan (1989) has noted how, in this class, laboratory benches were arranged in a tiered amphitheater around the lecturer's podium. Reproduced from E. C. Robins, *Technical School and College Building* (1887), plate 30, by permission of the Syndics of Cambridge University Library.

the head) was seen as separate from the work of the hands; the two activities could be performed by distinct individuals. Technicians frequently appear in the record only because they were blamed for mistakes. Their supposed independence from the reasoning processes that guided an investigation was a resource that could be deployed to account for failure.

Roger Stuewer (1985) has recounted a modern parallel. The lengthy controversy in the 1920s between groups in Cambridge and Vienna concerning the mechanisms of atomic disintegration was resolved without a damaging loss of status on either side when errors were located in the methods of scintillation counting used by female technicians. In other words, support staff bore the responsibility for worrying discrepancies in observations. This was found to be a more agreeable solution than blaming the physicists on either side, presumably at least in part because the technicians were not consulted and their judgment could more easily be doubted. As Achinstein and Hannaway note, in introducing Stuewer's account of the episode, "a division of labor had separated the eyes of the observers from the minds of the experimenters" (Achinstein and Hannaway 1985: x).

The history that separates the technicians in Vienna in the 1920s from the anonymous "operants" in Boyle's laboratory is unquestionably a long and complex one. The hierarchy of the division of labor has been formalized to some degree, but its precise configuration has varied with different assemblages of equipment and different forms of laboratory work. Simon Schaffer (1988) has traced systematic attempts to discipline laboratory personnel to the early nineteenth century, when methods were taken over from industrial management. He has studied the use of the "personal equation" to standardize observations by those hired as astronomical assistants in the period. The idea was to quantify the reaction times of individuals so as to be able to correct for each observer's errors in recording the times of transits, eclipses, and so forth. Use of the equation promised to make all observations reducible to a single standard. It took its place within a structure of discipline that sought to reduce laboratory practice to machinelike operations, performable by any practitioner interchangeably. The regime introduced by G.B. Airy at the Royal Observatory at Greenwich provided a model of factory-style discipline copied by the observatories at Paris, Brussels, Göttingen, Pulkovo, and elsewhere. Schaffer writes: "The observatory became a factory, if not a 'panopticon.' 'Mere' observers were relegated to the base of a hierarchy of management and vigilance, inspected by their superiors with as much concern as were the stars themselves" (1988: 119).

Peter Galison (1985) has shown how the management of personnel has been put into practice in high-energy physics in the last few decades, in connection with use of the bubble chamber. The apparatus, developed initially by Donald Glaser in the 1950s, was refined to yield a large number of photographs of particle physics events. Since only a very small proportion of these photographs would show significant new events, a data bottleneck was created. Luis Alvarez, at the University of California at Berkeley, responded to the challenge with a series of innovations in data-analysis technology and labor organization. He designed a system whereby relatively untrained technicians (usually women) would perform routinized operations to translate photographs of particle tracks into quantitative data that could be fed into computers. The physicists dealt only with the products of this process – intermediate stages had been delegated to machines and to unskilled and disciplined personnel. "Human engineering" had been employed, alongside the practices of conventional engineering, to make the production of new knowledge into a factory-like process. Alvarez's innovations in labor management have set the tone for much of the "big science" of recent years.

As Galison points out, attention to these questions of the organization of work in the laboratory adds another layer of analysis to the history of science conceived as theoretical development and experimental practice. He emphasizes that "The organization of specialized subgroups and the integration of engineers, programmers, and scanners was as much a component of the changing experimental physics of the late 1950s and the early 1960s as the bubble chambers themselves" (1985: 356). The different levels interact, moreover, at many points. New technologies and methods allow for the redistribution of skills between people and machines, including "deskilling" of human beings to make them work as part of ensembles of apparatus. The issue of the degree to which machine agency can be substituted for that of humans, raised by Collins, Latour, and others, becomes relevant at this point. The full history of the complexes of theoretical systems, material ensembles, and human actions that the laboratory has manifested will be a convoluted and a lengthy one – perhaps one for which traditional narrative historiographical forms will not be adequate (Galison 1988). It is, however, the history which the institution of the laboratory challenges us to write.

BEYOND THE LABORATORY WALLS

The laboratory always exists in relation to an external realm. This is true in several senses. First, the laboratory is impossible without external support; financial, material, and cultural resources must be gathered to maintain its activities. Second, the laboratory must be deliberately isolated from certain kinds of surrounding influences, not just the disturbing presence of unqualified visitors but also sources of physical interference. Cahan's (1989) account of the building of the Physikalisch-Technische Reichsanstalt particularly emphasizes the need to screen out disturbance from mechanical vibrations and electromagnetic flux. Lengthy political negotiations were conducted to try to prevent the Berlin streetcar company from running a tram line along the front of the building. Although ultimately unsuccessful, the negotiations delayed construction of the streetcar line long enough for measures to be developed to shield the equipment that would have been affected.

Third, and most obviously, laboratory-made knowledge does not remain there but emerges into the world outside. One very common way in which this is accomplished is through demonstrations of experimental effects in front of an audience of some kind. As Harry Collins has noted (1987a, 1988), a demonstration is essentially rather different from an original experiment; it is a rehearsed reiteration of an already achieved effect, in which the desired outcome is known in advance. It is accomplished through manipulation of material apparatus, together with the discursive packaging that conveys its meaning for the audience. It also requires the establishment of certain kinds of relations between demonstrator and audience, such that the audience knows what to expect and how to behave. The spatial relations and behavioral repertoires characteristic of the *theater* have frequently been exploited for this purpose, although reservations have sometimes been expressed about the appropriateness of the model.

The relationship between public experimental philosophy and the theater was explored already in the seventeenth century. Francis Bacon warned of the dangers of "the idols of the theater" in connection with experimental science. Established systems of philosophy, he argued, represented "worlds of their own creation after an unreal and scenic fashion" (1960: 49). Surrendering to the seductions of narrative simplicity and dramatic spectacle that characterize the theater, traditional philosophy subordinated critical curiosity to wonder. Although Bacon's aim was to free experimental inquiry from this taint, the idol was not as easily dislodged as he wished. The members of the early Royal Society continued to ponder the irreducible element of dramatic performance that attended public experimentation. While the wonder of assembled spectators was a potentially useful resource, it needed to be carefully circumscribed and managed. The Society would have preferred not to be confused, as Hobbes provocatively suggested, with those "who deal in exotic animals, which are not to be seen without payment" (quoted in Shapin and Schaffer 1985: 112). Christopher Wren, writing in 1663 about plans for a proposed visit by Charles II to the Society, posed the problem directly. On such an occasion, he wrote, "there ought to be something of pomp." But, on the other hand, "to produce knacks only, and things to raise wonder, such as . . . even jugglers abound with, will scarce become the gravity of the occasion" (quoted in ibid.: 31).

In the eighteenth century, much scientific activity moved into the settings characteristic of what Jürgen Habermas (1962/1989) has called "the emergent public sphere," including coffee shops, public houses, recreational resorts, and assembly rooms. In many of these settings, experimental demonstrations continued to invoke the model of the playhouse, with its ambivalent moral implications. Newtonian public lecturers such as John Harris, William Whiston, and John Theophilus Desaguliers tried hard to differentiate themselves from "vulgar projectors" and wonder-mongers (Stewart 1992). Other practitioners, such as the electrical therapist James Graham, were not so scrupulous and exploited the model of popular entertainment more shamelessly (Altick 1978; Stafford 1994). Controversy surfaced repeatedly about displays of electrical effects, as lecturers were accused of making a natural phenomenon appear miraculous, or of improperly appropriating a divine power (Schaffer 1983, 1993). Even the undemonstrative chemist Joseph Priestley was charged by a rival, Bryan Higgins, with making essentially commonplace phenomena appear "perfectly mysterious and surprizing to others" (Golinski 1992a: 89). It seemed as if the effort to display experimental knowledge publicly was inherently compromised by the allure of dramatic spectacle.

In Britain, the issues surrounding the theatrical form of public science took on a strongly political tone in the final decade of the eighteenth century. In a climate of intense political polarization after the outbreak of the French Revolution, public science was subjected to unprecedented strains that continued to shape it well into the following century. On the one hand, scientific showmanship reached across to broader audiences among the working classes, with a program that included mesmerism and phrenology. On the other hand, a more specialized experimental science emerged within newly formed institutions, exploiting technically advanced instrumentation. This specialist science was, nonetheless, still oriented toward a public audience – the upper and middle classes, who were viewed as the potential patrons of research (Cooter and Pumfrey 1994). Humphry Davy's carefully stage-managed and rhetorically crafted displays for wealthy audiences at the Royal Institution in London invoked a new model of the theater. His respectable, well-behaved audience pointed the way toward the disciplined population who attended university lectures in the nineteenth century (Forgan 1986: 102–103). However, Davy's practice retained certain signs of its eighteenth-century roots: the admission of women to lectures, for example, and the invitation of a select group to witness research experiments in a laboratory fitted up for display (Golinski 1992a: chap. 7).

Davy's successor at the Royal Institution, Michael Faraday, maintained a firmer demarcation between the places of private work and public demonstration. He made creative use of transitions between the two spaces to consolidate his experimental findings and to communicate them. David Gooding's impressive studies (1985a, 1985b) have shown how the movement from the laboratory to the theater strengthened Faraday's own understanding of the phenomena of electromagnetism and helped persuade other physicists that he was correct. Confirming Fleck's thesis on the effect of movement between "esoteric" and "exoteric" realms, Gooding shows Faraday working to magnify the phenomena, while masking the labor involved in their production, so as to make nature appear to speak directly to his audience. Gooding remarks: "To become accepted as part of scientists' experience [a natural] phenomenon must be transferred from the personal realm to the public domain, where it can be reproduced and witnessed by all. . . . Faraday was good at moving his discoveries from the personal domain of the contingent to the public forum of the demonstrable and self-evident" (Gooding 1985a: 105). Geoffrey Cantor has recently linked this careful management of the relations between private and public realms to Faraday's religious orientation as a member of the small Sandemanian sect. The Sandemanians typically practised a degree of aloofness from the everyday world, and, although Faraday was not a recluse, he carefully protected the domain of his private life and adopted a somewhat stylized persona when presenting himself in public (Cantor 1991b: 110–118, 151–154).

Very specific factors such as this no doubt shape the configurations of private and public spaces, and the relations between them, in particular instances of experimental work. Thomas A. Markus has given a detailed typology of lecture theaters, suggesting techniques for connecting the spatial arrangements of the buildings to the relations of power that they embody and exploit in the production of knowledge (1993: 229–244). The general model of the laboratory and the theater as two ideal types of space, with artifacts and representations making the transit from one to the other, seems applicable to many situations in which natural knowledge is produced. Shapin has written that "The career of experimental knowledge is the circulation between private and public spaces" (1988b: 400). Experimental science is constructed initially in secluded locations where the necessary resources are concentrated and protected, but it is only stabilized by the transfer of artifacts and representations to sites where they can be displayed to an audience.

Although this is indeed a widely useful model, it would be wrong to suggest that it covers all situations in which natural knowledge is constructed. The passage from the laboratory to the theater may be archetypal, but it is not universal. For one thing, many of the locations in which knowledge is rendered public do not conform to the model of the theater. The meeting places of scientific societies frequently invoked quite different associations in their architecture (classical temples or churches, for example) and thereby sought to elicit other than theatrical forms of behavior (Forgan 1986). Rudwick's discussion (1985: 18–27) of the parliamentary form of seating arrangements at the Geological Society of London in the early nineteenth century indicates how spatial forms borrowed from political institutions might be deployed to frame scientific debate. More fundamentally, we should also consider cases in which natural knowledge is not made through passage from private to public realms at all. In the remainder of this chapter, I shall discuss two of these: the *museum*, and the *fieldwork site*. The first is a setting in which natural knowledge is constructed in the very process of display itself, without that display making reference back to some anterior location or previous occasion of private experimental work. The museum presents what Markus calls "visible knowledge"; the things shown there are made known in the act of being displayed (1993: 171–212). Artifacts and natural objects displayed in museums have been gathered together from a variety of places, and they may be interpreted as signs of something else (for example, the world of nature) that is not viewed directly; but it is the act of showing that directly makes them known. This contrasts with theatrical displays or demonstrations, which are taken as reenactments of some knowledge-producing action that has previously occurred elsewhere. Barbara Stafford refers to this allusion to a prior, unseen realm as the "visible invisible" of the lecture or show (1994: 73–130).

A number of historians have recently urged that more serious attention be paid to the museum as a site for the production of natural knowledge. Its roots have been traced to the "cabinets of curiosities" maintained by early-modern virtuosi (Daston 1988, 1991; Findlen 1989, 1994; Impey and MacGregor 1985). There, in collections of astonishing profusion and heterogeneity, artificial and natural objects collected from far afield were displayed in apparently chaotic arrangements. Fossils, skeletons, stuffed animals, ancient coins, intricately carved cherry stones, relics, American Indian artifacts, gems, pictures, and books were among the things to be found in such collections. The criteria for inclusion were rarity, value, origin in some distant place, or a somewhat vaguely defined "curiosity." This last was also taken to be an attribute of the collector, so that the cabinet reflected his personal prestige and accomplishments. Maintaining a collection of curiosities and showing it to distinguished visitors was a way of displaying one's standing, especially in the status-conscious culture of the Renaissance court (Tribby 1992). The reference of the collection to the realm from which the objects had come – the natural world – was initially less direct, though Lorraine Daston (1988) has argued that the cabinets encouraged the attitude of particularistic nominalism that was an essential component of the new empirical sensibility of seventeenth-century natural philosophy.

As settings for courtly interaction and civil conversation, early-modern museums occupied a specific sociocultural niche. They could be said to be situated between private and public realms, but in a rather different

Figure 4. A Renaissance cabinet of curiosities: the kind of institution to which the roots of the modern museum have been traced. This is Ferante Imperato's museum in Naples, from the late sixteenth century. Frontispiece from *Dell'historia naturale di Ferrante Imperato libri XXVIII* (Naples, 1599). Reproduced by permission of the Syndics of Cambridge University Library.

way from the laboratory. Museums originated from the secluded studies or closets to which the Renaissance humanists retreated to cultivate the *vita contemplativa*. These were private, domestic, exclusively male spaces, at least according to the architectural model specified by Leon Battista Alberti in the early fifteenth century (Findlen 1989: 69). Later, however, they opened out into the world of courtly civility, as the entertainment of visitors became a prime function. Paula Findlen writes: "By the seventeenth century the museum had become more of a *galleria* than a *studio*: a space through which one passed, in contrast to the static principle of the spatially closed *studio*." A visitor to the collection of Ulisse Aldrovandi in Bologna in the late sixteenth century praised "his Theatre of nature, visited continuously by all of the scholars that pass through here" (Findlen 1989: 71; cf. 1994: 109–146). By the eighteenth century, to the extent that civil life had been opened up to women of the "polite" class, they too were to be found in museums.

It was in the eighteenth century that the museum became the prime site for construction of a specific form of scientific knowledge. Museums, and such analogous institutions as botanical gardens and mineralogical collections, became central to the sciences that Michel Foucault subsumed under the general label of "natural history" (1966/1970: chap. 5). Such disciplines as botany, zoology, and mineralogy were organized around the specific spatial relations characteristic of the museum. Objects were studied for their visible, surface features; they were isolated from one another (whether in the growing beds of botanical gardens or the finely divided drawers of mineralogical cabinets) and presented to scrutiny from a single point of view. Above all, *order* was displayed in the arrangement of individual specimens: Their rigorous placement in relation to one another made manifest the possibility of a classification that was thought to correspond to a "natural" order in the world. The heterogeneous, chaotic arrangements of the Renaissance and Baroque cabinets gave way to very deliberate groupings of items according to their membership in distinct genera and classes (Jardine, Secord, and Spary 1996).

Foucault has plausibly argued that the sciences of natural history underwent a fundamental change at the end of the eighteenth century with the emergence of such new disciplines as geology and biology (cf. Albury and Oldroyd 1977). Chemistry, which in the Enlightenment frequently adopted the methods of natural history, was also transformed at this time (Roberts 1991, 1993). The space of the museum continued, however, to be a resource of considerable importance in many of the sciences. Many new institutions were built in national capitals and provincial centers during the nineteenth century; and museums were routinely included at universities where the life sciences, earth sciences, or medicine were being taught. Natural history museums generally attempted to reproduce classificatory order in the arrangement of their rooms and display cases, as Louis Agassiz's plans for the zoology museum at Harvard show. John Pickstone (1993, 1994) has speculatively suggested that the museum assumed an entirely new importance in relation to a cluster of sciences that arose in the early nineteenth century. In such disciplines as comparative anatomy, pathology, and chemistry, the museum (especially when attached to an educational institution) became the setting for diagnosis of specimens by analysis into their constituent parts. Even engineering education made use of collections of machines organized according to their basic components.

Although its history in relation to the construction of scientific knowledge is only beginning to be written, it is apparent that the museum can claim a significant place on the map of locations in which science has been made. The museum comprises an enclosed setting, but one that can open out in various ways to the world beyond. It can be adapted to the tasks of education or popularization, but it can also serve as a site of research activity. Arrangement of its contents can signal various conceptions of the order that is believed to exist in the natural world and of the human relationship to it. Museums thus encode and shape particular configurations of knowledge; they display objects but they are never simply windows to the world beyond. The place in which the display occurs is crucial.

In museums – as in laboratories and such other sites as observatories, libraries, and hospitals – natural knowledge is constructed in a specifically designed and enclosed space. The fieldwork sciences, however, among which should be included ecology, geography, demography, anthropology, meteorology, geology, and climatology, are not restricted to such demarcated locations. The practitioners of these disciplines may sometimes work in offices, for example in universities, but they make their knowledge (at least partially) elsewhere. It might be thought, then, that for fieldwork subjects an analysis of science as a localized construction makes no sense, since their knowledge-producing practices are not bound to any delimited space.

A constructivist analysis of the fieldwork sciences is nonetheless beginning to emerge, although more work remains to be done to apply it historiographically (cf. Kuklick and Kohler 1996). It is clear that the kind of model that has been used for the laboratory will not work for subjects pursued beyond its walls; but the practices that have been observed in laboratory studies may also be found outside. And the category of spatiality may be very pertinent, albeit manifested rather differently in relation to fieldwork. Practitioners of the fieldwork sciences may be seen to be involved in constructing representations of their world, manipulating spatial relations so as to render the wider world accountable within scientific practices that are nonetheless, in substantial respects, local. In some cases, they may construct microcosms of conditions in the outside world for reproduction in the laboratory, in what Peter Galison and Alexi Assmus have called "mimetic experimentation," for example, that which originally produced the cloud-chamber as an attempt to mimic meteorological phenomena (Galison and Assmus 1989). More typically, what is brought back from the field is some visual representation of pertinent conditions in the location under study.

Analysis of the fieldwork disciplines in this way can learn a lot from Latour (1983, 1987: chap. 6). He stresses that the movement of the practitioner into the "field" is directed toward achieving a "translation." Methods are applied that yield a representation of some phenomenon that extends over a wide spatial sweep, and that representation is brought back for local use. The representation has the character of an "immutable mobile," a trace that is conveniently sized (generally smaller than the original) and fixed in some relatively permanent form. Specimens of animals or plants could be examples of these; assembled in a natural history museum or botanical garden, they serve as representatives of a distant fauna or flora. Other examples include maps, statistical tables, the results of questionnaires, photographs, anthropologists' field notes, readings of meteorological instruments, and so on. At a "center of calculation," where immutable mobiles are collected and processed, distant phenomena are brought closer. Only by bringing the world into the laboratory in this way, Latour argues, can scientific knowledge be made to encompass the world.

The various practices of *mapping* are obvious ways in which spatial relations are manipulated to create a locally usable representation of an extended space. Historians have begun to consider how mapmaking is embedded in localized practices. Maps have many different forms because they are created by adapting various representational practices to serve specific purposes. Particular visual renderings of extended space are produced for particular local uses.

Jacques Revel (1991) has considered the development of mapping techniques applied to the national territory of France. Although the nation was clearly associated with control of a certain territory by late medieval times, Revel points out that early attempts to accumulate knowledge of this territory did not take the form of maps. Inventory surveys of localities were the favored instrument for the state to assess tax revenues; these might be combined into statistical digests of regions or developed into discursive natural histories of specific places. Visual representations of space did exist – for example, the long scrolls on which a road route between two places was depicted in linear form, or the diagrammatic "world maps" that portrayed somewhat idealized cosmologies – but these techniques were not applied to the national territory.

The development of the visual arts in the Renaissance, in conjunction with the revival of classical geometry, yielded new techniques for the representation of spatial relations over a geographically significant area (cf. Alpers 1983: chap. 4). Renaissance geographers drew upon the *portolano* charts used by sailors to navigate along a coastline and upon the tables of latitudes and longitudes recorded by Ptolemy. Combining these with techniques from perspective and landscape art, they created maps in which geometrical relations between places on the ground were reproduced in reduced size on paper. As Revel comments, "Maps did not invent the sense of space, but they gave it a perceptive, conceptual, and technical form, which eventually became inseparable from 'spatiality' itself" (1991: 148). These maps were originally devised for display at court as affirmations of monarchical power over the territory. The king and his courtiers were privileged to occupy the point of view (which corresponded to no possible actual view) from which the whole kingdom could be encompassed: "The king could now sit in his chamber and 'without troubling himself greatly, see with his eye and touch with his finger' the expanse of the territory – without having to travel at all" (151).

During the eighteenth century, the geometrical grasp of the national territory was strengthened by developments in surveying techniques. Enlightenment geodesy was pioneered in France, but dramatic advances in instrumental engineering enabled the torch to pass to England in the second half of the century (Widmalm 1990). Collaboration between the two countries on the Paris–Greenwich triangulation in the years 1784 to 1788 put the new instruments and techniques to the test. By extending the reach of these technical practices, the space of national territories was brought within the scope of visual representation. An accurate portrayal of the geometrical outline of the state was the first priority in France; topographical detail was only filled in later (Revel 1991: 155–157).

We are now so accustomed to maps as means of representing geographical territory that their historical rootedness in specific local practices is easily forgotten. One way to reinforce the point is to compare the techniques of geodetic mapping with those used by geologists. Martin Rudwick (1976) has discussed how geologists, in the decades of the late eighteenth and early nineteenth centuries, developed a quite new relationship with the landscape and explored new ways of representing it visually. Their work was dependent upon the development of more sophisticated techniques for portraying topographical features in the same period. Once the visible features of surface landscape were shown by clearly understood conventions, geologists could adapt topographical maps to represent what could *not* be seen underneath.

As Rudwick points out (1976: 159), geological maps necessarily packed a high degree of theoretical interpretation into their visual form, because they were dedicated to showing what can only be inferred from surface evidence. The maps revealed the three-dimensional structure of rock strata and also indicated the metaphorically "deeper" level of causal change over time. The most common maps – those which displayed the distribution of rock types just beneath the surface of the earth – were thus complemented by two other kinds of diagram. One was the column, which purported to show the original order and thicknesses of strata prior to folding and erosion. The other was the section, an imagined exposure along a vertical cut through the landscape. All three forms of image embodied the results of reasoning that remained, to some degree, conjectural; they were continuously juxtaposed and adjusted with respect to one another in geologists' discourse (cf. Rudwick 1985: chap. 3).

Mapping is a paradigm example of the representational practices of the fieldwork sciences. The many different kinds of maps – geological, political, epidemiological, ecological, meteorological, and so on – comprise a large part of the product of those disciplines. By studying the conventions of visual representation drawn upon in mapmaking, the purposes various maps serve, and the contexts in which they are used, we can get a feel for the concrete practices of "translation" in which the fieldwork sciences are implicated. Maps can be read as more than transparent pictures of the world; placed in context, they reveal how representations of that world are constructed by adapting available conventions, and by bringing back from the field an image that can be subjected to scrutiny and debate. It is by these means that extended regions of space can be made accountable within the localized practices of science.

Although this analysis would need to be extended to cover the other practices in which the fieldwork sciences are involved, it does at least suggest that they are located in space in a rather different way from the laboratory sciences. Whereas the latter are sited in places where phenomena are produced by clusters of instrumentation and skilled personnel, the former require travel and the means to mobilize representations of extended regions of space. In an analysis which parallels that given here, Ophir and Shapin (1991), have drawn upon an earlier discussion by Foucault to distinguish "heterotopic" and "nonheterotopic" places for the construction of scientific knowledge. "Heterotopias" include laboratories along with such sites as libraries, clinics, and museums; there, objects of knowledge are constituted within epistemic spaces that are other than the space occupied by the site. Ophir and Shapin write:

> One space enclosed within the [heterotopic] site is always a segment of an encompassing social space, with which it is contiguous, from which its agents come and to which they return. The "other" space is the one in which the objects of a science appear. It is the space in which such entities as "laws," "cells," "genes," "particles," "atmospheric pressure," and "mental illness" are made manifest and represented. (1991: 14)

The sciences practised in such heterogeneous, "doubled" settings are distinguished from those pursued in "nonheterotopic" sites. Oceanography, geography, and the other fieldwork sciences make their objects of knowledge at places away from the routine workplaces of their practitioners. Their pursuit requires travel and the transportation of representational artifacts back from the field setting.

As I have indicated, Latour's work provides important resources for fleshing out this skeletal description of the "nonheterotopic" or fieldwork sciences. He shows how practices of standardization, inscription, translation, and so on, which comprise his solution to "the problem of construction," are *also* at the heart of the fieldwork sciences. This is notable notwithstanding Latour's tendency to prioritize the laboratory in his analysis. Although he declares that he is interested in how scientific practices reconfigure spatial relationships between the laboratory and the world outside, he assigns the laboratory itself pride of place in doing this. In his discussion of the work of Louis Pasteur, for example, Latour insists that he will not "use a model of analysis that respects the very boundary between micro- and macroscale, between inside and outside, that sciences are designed not to respect" (Latour 1983: 153). Yet it is Pasteur's laboratory that is the locus of the reversal of forces – the displacement of "actants" – that makes his success possible. Thus, as Jon Agar puts it, Latour "seems initially to dissolve the importance of the laboratory as a privileged site only to reinstate it with more vehemence" (Agar 1994: 28).

Latour's ideas can nonetheless be applied critically to sciences in which the laboratory is not crucial. This is, if anything, facilitated by his very loose definition of what a "laboratory" is: The kinds of practices that characterize it are equally to be found in such places as tax offices, military and corporate headquarters, and museums. This suggests that analysis of specific practices might be at least as illuminating as a focus on supposedly privileged places. Indeed, Michael Lynch (1991) has argued that science does not simply occupy certain pre-given places; rather, it constitutes spaces of action (which he calls "topical contextures") through certain practices and instrumentalities. These may, of course, extend beyond the laboratory, though whether they could dissolve the distinction between inside and outside, as Latour claims, remains unclear. A more scrupulous and discriminating analysis would have to pay attention both to the reconfiguration of spatial relations that investigative practices accomplish and to the ways they are laid out across space defined in other terms (architectural, geographical, ecological, and so on). We need to look closely at the boundaries that are crossed and those that are maintained, at the juxtapositions that are made and the distances that are preserved.

The laboratory has earned its status by the manipulation of relations between inside and outside, by using representational means to alter the magnitude of entities, by bringing into proximity objects that are initially distant from one another, and so on. These practices have made it a privileged place for the construction of natural knowledge. But it owes its success also to an ability to make the knowledge constructed in the laboratory travel beyond its walls. And this ability is connected, in terms of the practices involved, with the success of the sciences that study the world beyond the laboratory. Travel, translation, mapping, and the other techniques used for mobilizing traces of the spatially extended world are fundamental to the success of the fieldwork sciences *and* to the construction of laboratory science in the public realm.

4

Speaking for Nature

[W]e should not imagine that the world presents us with a legible face, leaving us merely to decipher it; it does not work hand in glove with what we already know; there is no prediscursive fate disposing the word in our favor. We must conceive discourse as a violence that we do to things, or, at all events, as a practice we impose upon them; it is in this practice that the events of discourse find the principle of their regularity.

Michel Foucault, "The Discourse on Language" (1971/1976: 229)

THE OPEN HAND

Constructivist studies have been built upon the supposition that science can be understood by investigating its observable practices. Repeatedly, in these studies, scientific practice has been shown to involve a great deal of verbal and written communication. A lot of what scientists can be observed to do is linguistic behavior. They converse with one another as they work, communicating the details of techniques and observations. They spend long hours drafting and redrafting grant applications. They read at length in the relevant literature, before composing, with great care, the papers in which results are reported. When they come together in scientific institutions, they participate in further communicative acts, such as delivering lectures or commemorative addresses, or debating the merits of one another's work.

Of course, not all scientific practice is discursive – at least, not purely discursive. The work of manipulating material objects also deserves scrutiny. And some communicative practices are nonverbal, particularly the production and circulation of images, such as graphs, photographs, diagrams, and the "inscriptions" produced by various kinds of instruments. (The manipulative and representational aspects of scientific practice will be discussed in Chapter 5.) Nonetheless, the discursive dimension of science is evidently of considerable importance. Scientists are very articulate language users; they live much of their lives in a world of words. Nor can constructivist analysis afford to dismiss the linguistic aspect of science as if it were purely epiphenomenal. Sometimes scientists insist that their language is of importance only for what it refers to:

103

"nature." But those who look at science from the outside, and who attempt to place it in its context, are bound to take more seriously the linguistic actions to which so much effort is devoted. The question then becomes, how is the "reality effect" achieved by scientific discourse? How do scientists place themselves in a position to be accepted as speaking for nature?

This accomplishment can be identified as a matter of persuasion. Scientific language works to persuade its audiences that they can read *through* it to apprehend nature. To understand how this is achieved, resort has been made to the discipline concerned with effecting and analyzing persuasion, namely *rhetoric*. Sociologists have been discussing the rhetorical function of lectures, research papers, and textbooks, sometimes in rather general terms, sometimes making use of specific resources from the rhetorical tradition. Historians have also considered how persuasive scientific discourse of various kinds has been created in specific cultural contexts. And rhetoricians themselves, perceiving that their subject has gained renewed fashionability after centuries of neglect, have applied their techniques to scientific writings. The "rhetoric of science" is now well established as a subfield of interdisciplinary science studies. (See, for example: Bazerman 1988; Dillon 1991; Gross 1990; Montgomery 1996; Nelson, Megill, and McCloskey 1987; Pera 1988; Pera and Shea 1991; Prelli 1989a; Weimar 1977.)

Rhetoric itself has, of course, been around very much longer than constructivism. As a discipline that teaches how to speak effectively, it dates back to the Sophists of the fifth century B.C. Plato famously attacked rhetoric on behalf of the dialectical method of arriving at the truth advocated by his mentor, Socrates. Since Plato's onslaught, reiterated denigration of rhetoric has been common in the Western tradition. It has customarily been denounced as an illegitimate technique for insinuating claims that have no intrinsic worth. Language that is viewed as having no reference to reality or correspondence with truth is frequently castigated as "mere rhetoric." Plato's denunciation of rhetoric in the name of philosophy established an antagonism between the two disciplines that recurred frequently during the following centuries (Vickers 1988: chap. 3). The uneasy rivalry persisted, notwithstanding Aristotle's reassertion of the value of rhetoric, which contributed significantly to keeping its tradition alive in Western culture. Aristotle elevated rhetoric to a level with dialectic as an equally valid means of conveying opinion, but he set philosophical reasoning above both of these, since it "starts with universal or necessary principles and proceeds to universal and necessary conclusions" (quoted in ibid.: 161). Rhetoric, then, was valuable provided it was firmly subordinated to philosophy, which enjoyed the exclusive prerogative of revealing truth.

Aristotle's exposition of rhetoric nonetheless provided a crucial source, along with the works of the Roman authors Cicero and Quintilian, for the revival of the ancient art among the humanists of the Renaissance. Rhetoric was represented emblematically as an open hand, in contrast to the closed fist of logic; the implication was that it yielded the means to set out an argument openly but did not compel assent (Howell 1961). This tolerant attitude toward rhetoric did not, however, survive unchallenged in the seventeenth century. The new philosophy of that period was vehemently antirhetorical. Francis Bacon's criticism of the "idols of the marketplace" set the tone for repeated denunciations of the deceptive allurements of language and attempts to forge a "plain" style that would be purely descriptive. The numerous schemes for artificial and universal languages sought to free discourse from the distortions of figural usage, so that each thing would have just one name (Slaughter 1982). Thomas Sprat claimed, on behalf of the early Royal Society, that their discourse utilized a pared-down, unornamented style, "to return back to the primitive purity and shortness when men deliver'd so many *things* almost in an equal number of *words*" (Sprat 1667/1958: 62). In the philosophical work of John Locke we find the culmination of this denigration of rhetoric as the obstacle to effective linguistic communication, a "powerful instrument of Error and Deceit":

> But yet, if we would speak of Things as they are, we must allow, that all the Art of Rhetorick, besides Order and Clearness, all the artificial and figurative application of Words Eloquence hath invented, are for nothing else but to insinuate wrong *Ideas*, move the Passions, and thereby mislead the Judgment; and so indeed are perfect cheat. (Locke 1689/1975: 508)

From the standpoint of its contemporary revival, we can discern that the vigorous denunciations of rhetoric by Sprat and Locke were themselves rhetorically crafted. Experimental philosophers frequently castigated figural language by helping themselves to colorful and highly charged metaphors, including many that identified rhetoric with the despised attributes of the female gender (vanity, cosmetics, harlotry, and so forth). The "plain" style, furthermore, can itself be viewed as a species of rhetoric. Even Locke was not willing to dispense with the rhetorical instruments of "Order and Clearness" for getting his message across. The presence of crucial rhetorical and metaphorical elements in his writings undermines his declared aim of freeing language from rhetoric to enable it to serve as the transparent medium of ideas (Bennington 1987).

One of the consequences, then, of a heightened awareness of rhetoric is to enable us to expose modern scientific writing as rhetorical *malgré lui*. It opens the door to a detailed analysis of the persistence of certain figures and tropes from the rhetorical tradition in scientific texts. It also, however, carries a significant ideological charge. The assertion that scientific discourse is concerned with the banal matter of persuasion – that truth does not shine through language as clear as glass, but that opinion must be moved by means that may be similar to those used in politics and the law – is frequently seen as a challenge to science's claims to truth and objectivity. Science's high epistemic profile in our culture is bound up with the notion that it produces knowledge of a higher degree of certainty than that yielded by purely human processes of persuasion. To talk about science as rhetoric, it is sometimes feared, is to consign scientific deliberation to the realm of irrationalism, religious conversion, propaganda, and "mob rule."

This is surely an overreaction; but it is perhaps an understandable response to a sometimes overpressed and underdeveloped argument. To announce simply that "science is rhetoric" is to tell us nothing about *how* scientific discourse achieves its persuasive effects. Unless we can discriminate the specific rhetorical techniques used in science, we risk being taken as simply attacking its epistemological claims. As Steve Fuller has noted, "The realization that 'everything is rhetorical' has much the same ring as learning that one has been speaking prose all of one's life. Nothing much changes . . ." (Fuller 1993: xii). To go further requires a more discriminating analysis of particular discursive resources and how they have been used. Short of this, we have no more than a slogan: another stick with which to beat epistemological pretensions to objectivity and realism (should that be one's aim), but one that ironically relies for its effect upon the pejorative connotations of rhetoric that rhetoricians usually claim to have transcended. As Greg Myers pointedly asks, paraphrasing Michael Lynch, " 'Rhetoric as opposed to what?' The claim that scientists are using rhetoric is only interesting as an ironic debunking of the assumption that their discourse is especially 'objective.' Once one grants that this objectivity is something they create in their work, the claim that everything is rhetoric has little meaning" (Myers 1990: 31).

Rhetoricians themselves have developed a variety of responses to this dilemma. Some have sought to show that the rhetorical tradition can yield techniques of sufficient subtlety to intervene productively in scientific debates. Lawrence Prelli (1989a) adapts the rhetorical procedure of *stasis* to provide protocols for articulating what is at issue in any controversy. By following this procedure, Prelli claims, scientists may exhaustively catalogue the issues that divide them as a step toward resolving their disputes. He proposes: "The system of *stasis* analysis for scientific rhetoric not only allows understanding the strategic choices rhetors did make, but also consideration of alternative possible choices they could have made" (Prelli 1989a: 176). More ambitiously, Steve Fuller (1993) proposes to use rhetorical techniques of "interpenetration" to defuse conflict between alternative disciplinary views of a wide range of controversial questions, and to point the way toward more fertile interdisciplinary communication. Far from undermining science's claims to knowledge, these analysts try to show that rhetoric can help scientific practitioners resolve the discursive difficulties that hinder their progress.

For historians, on the other hand, rhetorical analysis is likely to establish its utility to the degree that it can be yoked to the aims and methods of social or cultural history. Broadly speaking, we may identify three basic categories that rhetoric offers for the analysis of scientific discourse in its historical context: *convention*, *audience*, and *situation*. The first denotes the shaping of discourse by certain formal protocols that provide resources and establish constraints affecting the speaker or writer in a particular setting. The second category refers to the notion that all discourse embeds within it a certain orientation toward prospective listeners or readers. Speech and writing are aimed at audiences, whose interests they are framed to enroll and whose potential objections they seek preemptively to answer. Even if nobody actually reads a text, an "implied reader" is nonetheless invoked, whose characteristics are specified by the way in which it is written. Situation, finally, links the adoption of certain discursive conventions to the setting in which rhetor and audience are located and in which a certain range of speech acts is deemed appropriate. As Prelli points out (1989a: 21–28), this grasp of the contextuality of discourse has been part of the rhetorical tradition since Aristotle.

Convention, here, covers more than just genre, the species by reference to which a particular piece of discourse can be classified as a commemorative address, an undergraduate lecture, a grant proposal, a paper in *Nature*, a work of popularization, or whatever. Conventional norms can also be seen to shape discourse at many other levels. Rhetorical analysis can also consider figures of speech, choice of vocabulary, style and syntax, arrangement of subject matter, the "ethos" or projected persona of the author or speaker, the ways in which the audience's emotions are appealed to, the functions of humor, and so on. Everything that might be classed among the "formal" aspects of discourse can be included under this heading. "Form" is, of course, traditionally contrasted with "content"; but rhetorical analysis tends to be – at least to some extent – subversive of this dichotomy. Some rhetorical analysis stops short of engaging with the subject matter of science, as scientists themselves would understand it. But others take a more ambitious view of the potential scope of the method, asserting that form (understood sufficiently broadly) is exhaustive of what has traditionally been taken as the content of scientific discourse or arguing that the dichotomy between form and content must itself be deconstructed in light of the rhetorical perspective.

Let us consider, then, how the categories of convention, audience, and situation have been put to work in historical studies. It seems appropriate to begin with research on the seventeenth century, given the strongly voiced antipathy to rhetoric in that period. Not surprisingly, rhetorical analysis has not been content to echo the self-appraisals of the early writers on experimental philosophy. The "plain" style has been shown to have been anything but artless and simple; it has emerged, rather, as a highly refined instrument in which conventional resources were mobilized to construct persuasive construals of experienced phenomena for quite specific audiences.

Steven Shapin's (1984) treatment of the experimental writings of Robert Boyle has set the tone for much of this recent discussion. Shapin does not, in fact, use the term "rhetoric" at all, preferring his own coinage, "literary technology." The neologism captures the sense that Boyle's language was an instrumental resource, a tool he could mold and wield to serve his persuasive purposes. It also suggests a close connection with Boyle's material practice, the construction and manipulation of the air pump being portrayed as directed toward the same persuasive ends as his writing. In addition, by avoiding explicit talk of rhetoric, Shapin perhaps seeks to evade questions about the relationship between the rhetorical and the referential functions of language. He can display how Boyle's writing works to secure conviction without being obliged to answer the quite distinct question of the degree to which the author was describing what actually happened in his laboratory.

Shapin's analysis is nonetheless framed in terms that have elsewhere been identified as central to the rhetorical tradition. He remarks, "I shall attempt to display the conventional status of specific ways of speaking about nature and natural knowledge, and I shall examine the historical circumstances in which these ways of speaking were institutionalized" (Shapin 1984: 481). The notion of language as the persuasive deployment of conventional resources in a specific setting has long been of fundamental importance to rhetorical theory (cf. Prelli 1989a: 6–7, 11–32). Shapin goes on to specify how certain stylistic features of Boyle's texts were tailored to the purpose of conveying a convincing account of the experiments described. The prolixity of Boyle's prose, his relentless description of circumstantial details (even of failed experiments), and the naturalistic pictures of his apparatus, were all means toward persuading the reader of the factual status of his accounts. "Appropriate moral postures" were also adopted to communicate the persona or "ethos" of the author, including an explicit candidness about failures, a display of modesty in advancing opinions, and a reluctance to engage in disputes. In addition, the factual credentials of the accounts were buttressed by the testimonies of reliable witnesses, preferably men of gentle birth, whose names and rank were sometimes recorded.

Shapin connects these stylistic innovations to the kind of audience that Boyle's writings invoked. In line with the traditional rhetorical deployment of the category of audience, Shapin does not discuss the reactions of Boyle's *actual* readers; his concern is rather with the function of the texts in relation to a projected audience that they themselves worked to realize. Boyle's writings are thus said to have instantiated a number of the features that he proposed as foundational norms for the community of experimental philosophers: a clear demarcation between descriptive fact and theoretical interpretation; practices of collective witnessing; experimental demonstrations in public spaces; and the regulation of disputes by gentlemanly good manners. In these respects, "Boyle was endeavouring to constitute himself as a reliable purveyor of experimental testimony and to offer conventions by means of which others could do likewise" (Shapin 1984: 493).

To these two rhetorical themes – formal conventions and constructed audience – Shapin adds a third: the situation in which discourse occurs. To a certain degree in his 1984 paper, and more substantially in the book written with Simon Schaffer (1985), he links Boyle's choices of stylistic resources to the setting in which philosophical discourse was to function, one in which social consensus had to be reforged after the violent disruptions of civil wars and revolution. The intellectual elite of Restoration England was strongly inclined toward reconciliation of the diverse theological and philosophical positions that had recently proved so catastrophically divisive. Boyle's discursive innovations were offered as an appropriate response to this situation. By carefully maintaining the distinction between experienced fact and theoretical interpretation, Boyle proposed to relocate agreement around the results of experiments while allowing latitude of philosophical doctrine. Assent would be guaranteed to witnessed "matters of fact" while dissent would be limited to the realm of personal "opinion." In this way, the "calm space that experimental philosophy was to inhabit would be created and maintained through the deployment within the moral community of appropriate linguistic practices" (Shapin 1984: 507).

Utilization of formal conventions in discourse, the working of texts to create and discipline their audiences, and the appropriateness of discursive choices to particular settings are all themes that can be extended from rhetorical analysis per se into constructivist historical studies. Shapin provides a model of how this might be done with his demonstration that Boyle's experimental narratives both exemplified and advocated a specific set of choices concerning discursive form. James Paradis (1987) has also suggested that Boyle's accounts drew upon the rhetorical tradition of "essays" developed by the sixteenth-century humanist Michel de Montaigne. Loosely structured and digressive, each essay was framed as an autobiographical account of experience and accorded a place in a temporal process of investigation to which the reader was assigned the role of witness. Boyle's narratives, however, departed from the tradition of the humanist essay to the extent that they shifted attention from the internal, psychological world of Montaigne's reflections to the external, material world of experimental manipulations. "In doing so, [Boyle] transformed the self, a unique expressive intelligence in Montaigne's essay, into a passive instrument of observation, reporting on self-demonstrated material truths" (Paradis 1987: 60).

The narrative essay was not, of course, the only model of writing available to the experimental philosophers of the seventeenth century. Another was the didactic textbook, which was traceable to other trends within the broad humanist movement, specifically the revival of Aristotelianism and the sixteenth-century reforms of pedagogical methods inspired by Petrus Ramus (Schmitt 1973; Ong 1958). Owen Hannaway (1975) has traced the use of conventions of systematic exposition in Andreas Libavius's textbook, *Alchemia*, published at Frankfurt in 1597. For Libavius, the textual methods of Ramus, including the branching diagrams that showed how to subdivide topics into multiple dichotomies, provided an exhaustive articulation of the subject matter of the science and made it easier to memorize. They also had the advantage, from Libavius's point of view, of divorcing chemistry from the undesirable trappings of Paracelsian mysticism.

A further model for philosophical discourse was provided by the *dialogue*, of which the humanists found examples in the writings of Plato, Cicero, and other classical authors. As Greg Myers has pointed out in a recent treatment of the genre (1992), a dialogue is formally configured by its author with the devices of a recorded conversation, including digressions, interruptions, changes of subject, and so on. The purpose is generally to dramatize a process of learning or the achievement of consensus among speakers who originally hold divergent positions. Thus, Shapin suggests that Boyle used the form of dialogue in his work *The Sceptical Chymist* (1661) to demonstrate how agreement might be reached through civil conversation concerning matters of fact (1984: 503–504). This provided a model of the appropriate verbal and gestural conduct through which truth might be expected to emerge in philosophical intercourse.

In presenting a model of proper discourse, the dialogue mobilizes the fiction that the author is simply representing a real (or at least a plausible) conversation. Of course, as Myers points out, the author never in fact surrenders control. Dialogues do not resemble real conversations very consistently; they tend to merge into expositions of the author's own point of view. Nonetheless, the literary invocation of a dialogue serves to set the text at some distance from the persona and opinions of the author. An author may therefore use dialogue to advance an opinion while simultaneously stopping short of explicit avowal of it. Shapin notes this as one implication of Boyle's use of the form. For Galileo, who preceded Boyle in the use of dialogues, they offered rhetorical opportunities to explicate the Copernican cosmology while simultaneously disavowing actual belief in it (Cantor 1989: 167–173). Nicholas Jardine has argued (1991b) that this technique of authorial distancing might have seemed to Galileo an effective means to circumvent the church authorities' 1616 prohibition on belief in the Copernican system. Of course, as Galileo discovered when summoned to Rome for trial by the Inquisition in 1633, he sadly miscalculated the degree to which the authorities were willing to read his *Dialogues Concerning the Two Great World Systems* (1632) in this charitable way.

Today, the dialogue is a very minor genre of scientific writing, relegated to a marginal position as a means of popularization. Most scientific texts are concerned with reporting observations or the results of experiments to an audience of the author's peers. The genre of the research paper in the scientific journal has been developed to serve this purpose. Peter Dear has devoted a series of studies (1985, 1987, 1991b, 1992, 1995b) to exploring the roots of the experimental report in the seventeenth century. He stresses that the discursive conventions utilized for this kind of writing were not, by any means, readily available in the early part of the century, at least not within the scholarly tradition of writing on natural philosophy. The systematic treatises typically produced within that tradition contained no experimental reports as we would now recognize them. The crucial shift, Dear argues, was from a construal of "experience" as the summation of many commonplace events (not necessarily specified with respect to place and time) to the construction of detailed descriptions of a single circumstantiated and witnessed event. According to the former model, Aristotle was an authority because his texts contained the epitome of much experience of the natural world; according to the latter, what was required was to rebuild experience piecemeal from its foundations in properly warranted reports of discrete events. Like Shapin and Schaffer, Dear sees the institutionalization of the new discursive forms as preeminently the work of the Royal Society:

> When a Fellow of the Royal Society made a contribution to knowledge, he did so by reporting an experience. That experience differed in important respects from the definition informing scholastic practice; rather than being a generalized statement about how some aspect of the world *behaves*, it was instead a report of how, in one instance, the world had *behaved*. (Dear 1985: 152)

Dear does not follow Shapin and Schaffer in linking the rhetorical innovations of the Society to its setting amid the political crises of seventeenth-century England. Instead, he focuses on the roots of new discursive forms in deliberations about epistemology prior to the foundation of the Society, among Jesuit mathematicians for example. The two accounts are nonetheless complementary rather than contradictory. The rhetorical focus upon situation specifies that discourse is adapted to the setting in which it is produced, but it does not stipulate any particular social configuration or boundaries for that setting. And the emphasis on the conventionality of rhetorical forms makes it clear that literary traditions may make available resources from quite distant contexts for local application. To point to the political climate of Restoration England as a context for rhetorical innovation is not, therefore, to exclude the relevance of already established traditions of epistemological debate and discursive reform. Rhetorical analysis need not be bound to a narrow view of what the relevant context of discourse is.

There are, however, potential sources of tension between some versions of rhetorical analysis and constructivist history. They emerge, for example, from a study of the subsequent rhetorical development of the scientific research paper. Charles Bazerman's *Shaping Written Knowledge* (1988) is the most sustained treatment of how experimental reports have developed from the communications of the early Royal Society to the journal articles of today. He identifies the experimental report as a distinct genre of text, one that has assumed fundamental importance in empirical science. As a genre, the report is both a product of social interaction among researchers and a leading element of the context in which scientific discourse is produced. Scientific writers may proceed without much conscious awareness of rhetorical conventions, but this does not mean that such conventions are absent. Rather, they continue to exert substantial, albeit largely hidden, influence over the verbal expression of experimental facts. Bazerman adopts the classic stance of the rhetorician in specifying the aims of his enterprise: to identify and explicate the conventions that govern discursive production in experimental science in order to teach practitioners how to write and speak more effectively (1988: 3–17).

Bazerman does not attempt a comprehensive historical survey of the experimental report, but he does excavate various layers of its archaeology. Three chapters deal with the early Royal Society, three with twentieth-century physics, and two with the contemporary social sciences. Problems arise, not with the selectivity of the coverage as such, but with the overall schema of development that these episodes are taken to typify. Restricting himself, for example, to sampling the contents of the Royal Society's *Philosophical Transactions* at regular intervals in the period 1665–1800, Bazerman tends to assimilate his findings to rather vaguely characterized general trends. These usually take the form of uniform development toward institutionalization of the norms that Robert Merton identified in the modern scientific community. Bazerman sometimes adopts explicitly teleological language to describe this process, claiming, for example, that, "throughout the period, the increasingly expressed awareness of possible variables seems to reach toward an unexpressed concept of controls" (1988: 71). More generally, experimental writers are seen as making steady and unidirectional progress toward such modern ideals as impersonality, methodological caution, and maintenance of a systematic dialogue between experiment and interpretation. Bazerman summarizes:

> Commitment to organized criticism, communalism, universalism, and objectivity allow individuals to absorb individual strains, conflicts, and violations in the name of the communal endeavor. . . . [T]he general thrust of the development of the communication system of science has been to structure science in much the terms described by Merton. (1988: 148)

An eagerness to fit everything into a pattern of progress toward modern institutionalized norms colors Bazerman's identification of all of the trends he describes. It also masks the lack of any specific stipulations about historical causation. Rather than specifying cause and effect, Bazerman usually prefers to have things both ways. For example, certain readers of the early *Philosophical Transactions* are said to have already acquired a "professional identity" as "serious natural scientists," and yet that identity is also said to have been conferred upon individuals by publication in the journal (1988: 135, 138). In general, as Dear has noted (1988; cf. Bazerman 1987), the Mertonian model erects an ahistorical ideal of the "scientist" as the goal of a teleologically depicted process. Because Bazerman assimilates his historical evidence to such goal-directed trends, he readily moves between comments on the style of historical texts and remarks on the functional value of such stylistic conventions for the development of the modern scientific community. The effect is to produce a survey that is arguably not a history at all, but (as Dear puts it), "an account of the 'evolution,' in a Larmarckian, teleological sense, of the structure of *modern* scientific communities – characterized in Mertonian fashion" (Dear 1988: 275).

This tendency to lapse into teleological explanations is one evident weakness of an identification of rhetorical conventions with functional norms. The modern ideals appear to exert a seductive attraction, past innovations are interpreted as functional adaptations toward the goal of the present, and historical change is deprived of any more substantial causal forces. Another problem concerns the lack of attention to instances of controversy, when quite distinct rhetorical conventions can be invoked by different individuals or groups within the scientific community, frequently in connection with different models of scientific practice. Bazerman tends to assume that rhetorical conventions will be shared by all "scientists" as such, with the implication that they will always be available to provide a shared basis for the resolution of disputes. Constructivist studies of controversies have raised severe doubts about this happily consensual picture. Bazerman's restriction of his analysis to what would usually be recognized as the "stylistic" elements of the texts he discusses may also reflect a Mertonian unwillingness to engage with the content of scientific discourse – an unwillingness that constructivism, following in the wake of the controversy studies, has sought to overcome. These deficiencies weaken Bazerman's claim to have given a plausible historical account of the development of the experimental article and call into question his announced aim of reconciling Mertonian and constructivist approaches (1988: 129).

Scrutiny of controversial episodes can help to bring rhetorical conventions down from the lofty heights of supervening norms and link them to other elements of practice. As Prelli (1989b) has pointed out, in a rather differently formulated version of rhetorical analysis, the scientific "ethos" cannot be regarded as a given. He notes that Merton's attempt to specify "binding institutional norms that constrain the behavior of scientists" has been undercut by other sociologists' descriptions of scientists resorting to "counter-norms." In certain situations, scientific practitioners tend to favor values that are just the opposite of the Mertonian ideals. They may base judgments of others' work on assessments of personal attributes, for example, or they may endorse claims to individual property rights over discoveries. Prelli demonstrates how writers on both sides of a controversy concerning whether apes could learn the elements of human sign language deployed a range of sometimes contradictory normative claims regarding the qualifications and behavior of their opponents. In view of this, he proposes that neither norms nor counter-norms should be regarded as regulative principles; rather, they serve as rhetorical resources deployed in discursive negotiations in which factual claims and questions of proper scientific procedure are simultaneously at issue. To this extent, "scientific *ethos* is not given; it is constructed rhetorically" (Prelli 1989b: 49; cf. Mulkay 1979: 71–72).

Studies such as Prelli's remind us how wide-ranging arguments about factual claims may become. Disputes about whether apes are capable of human language can broaden to encompass issues of the competence of investigators, the propriety of their methods, or the appropriateness of different forms of publication. Ultimately at issue, we might say, in any sustained controversy, is how science should be carried on. In these situations, rhetorical conventions may well *not* provide a basis for consensus; instead they may themselves be at issue. Different discursive forms may be exploited, and arguments for their appropriateness voiced, to try to gain persuasive leverage in debate, but their legitimacy itself can be brought into question. When rhetorical forms themselves become subject to dispute, they are shown to be part of scientific practice, not external to it.

This point can be given substance by considering the dispute that followed Isaac Newton's publication of his paper "A New Theory about Light and Colors" in the *Philosophical Transactions* in 1672. Bazerman discusses the episode in an illuminating chapter (1988: chap. 4) in which he analyzes how Newton modified the stylistic form of his doctrine in a series of documents, from his private notebooks of the mid-1660s to the first book of the *Opticks*, published in 1704. He recognizes the importance of the controversy in which Newton found himself engaged with English Jesuits resident abroad and with Robert Hooke at home. But his analysis is very largely framed from Newton's own point of view. From the outset, it ascribes a fully formulated doctrine to Newton concerning the composition of white light from the colored rays that can be separated by refraction through a glass prism; and it postulates a series of rhetorical choices made by him to explain this doctrine to an uncomprehending readership. The analysis is subtle, but it is entirely confined to issues of expository style. Newton is seen as having successfully responded to the rhetorical challenges posed by his readers' inability to understand what he was claiming.

A more rigorously constructivist analysis of the controversy has been given by Simon Schaffer (1989). He sets the dispute within a framework, not of consensual norms shared by all "scientists," but of competing traditions. These are, on the one hand, the experimental natural philosophy pioneered by Boyle and the early Royal Society, and, on the other, the mathematical sciences whose range Newton hoped to extend to cover the subject of colors. Each tradition had its own practices of experimental manipulation and its own conventions for writing texts. Newton's initial framing of his claims in the 1672 paper as a narrative of discovery was evidently aimed at satisfying the conventional expectations of readers of the *Philosophical Transactions*; but he abruptly challenged those expectations by shifting, in the second half of the paper, from a description of his experiments to a programmatic enunciation of the propositions of his "doctrine." As Zev Bechler had already pointed out, in a very perceptive discussion of the dispute, this meant that "the problem of legitimate modes of writing science . . . played no small part in initially giving rise to the controversies and in helping maintain them to their unsatisfactory end" (Bechler 1974: 115).

Schaffer's analysis proceeds to display how many things were called into question in the ensuing debate. Although he emphasizes the dimension of material practice, particularly the question of the suitability for refraction of different kinds of glass prisms, he indicates that the propriety of certain discursive forms was simultaneously at issue. Most importantly, Schaffer shows how the designation of the very "content" of Newton's claims was controversial. The way the line was drawn between experimental "fact" and interpretive "hypothesis" was itself contested. The labeling of certain experiments as "crucial," the nomination of two experiments performed on separate occasions as essentially "the same," and the articulation of what had in fact been "shown" by an experiment

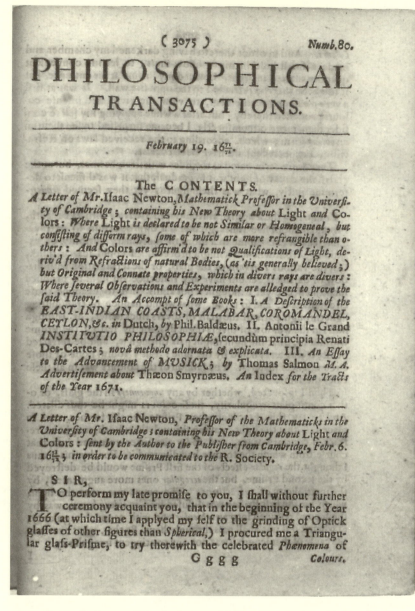

(3075)   *Numb.* 80.

PHILOSOPHICAL
TRANSACTIONS.

February 19. 16$\frac{71}{72}$.

The CONTENTS.

A Letter of Mr. Isaac Newton, *Mathematick Profeſſor in the Univerſity of Cambridge ; containing his New Theory about* Light *and* Colors : *Where* Light *is declared to be not Similar or Homogeneal, but conſiſting of difform rays, ſome of which are more refrangible than others :* And Colors *are affirm'd to be not Qualifications of Light, deriv'd from Refractions of natural Bodies, (as 'tis generally believed;) but Original and Connate properties, which in divers rays are divers : Where ſeveral Obſervations and Experiments are alledged to prove the ſaid Theory. An Accompt of ſome Books :* I. *A Deſcription of the* EAST-INDIAN COASTS, MALABAR, COROMANDEL, CEYLON, &c. *in* Dutch, *by* Phil. Baldæus. II. Antonii le Grand INSTITUTIO PHILOSOPHIÆ, *ſecundùm principia* Renati Des-Cartes ; *novâ methodo adornata & explicata.* III. *An Eſſay to the Advancement of* MUSICK ; *by* Thomas Salmon *M. A. Advertiſement about* Theon Smyrnæus. *An* Index *for the Tracts of the Year* 1671.

A Letter of Mr. Isaac Newton, *Profeſſor of the Mathematicks in the Univerſity of Cambridge ; containing his New Theory about* Light *and* Colors : *ſent by the Author to the Publiſher from* Cambridge, *Febr.* 6. 16$\frac{71}{72}$; *in order to be communicated to the* R. Society.

SIR,

TO perform my late promiſe to you, I ſhall without further ceremony acquaint you, that in the beginning of the Year 1666 (at which time I applyed my ſelf to the grinding of Optick glaſſes of other figures than *Spherical*,) I procured me a Triangular glaſs-Priſme, to try therewith the celebrated *Phænomena* of

                            Gggg   *Colours.*

Figure 5. The first page of Newton's paper "A New Theory about Light and Colors," from the *Philosophical Transactions* 6, no. 80 (19 February 1672): 3075–3087. Reproduced by permission of the Syndics of Cambridge University Library.

and what had been claimed as its implications were all open to challenge. Participants in the controversy retained considerable flexibility in respecifying what it was they had done on a particular occasion, or what they had asserted. For example, in June 1672, Hooke

> told the Royal Society's President that this trial might prove that 'colourd Radiations' maintain fixed refrangibilities: it did *not* prove what Hooke claimed Newton wanted to prove, that there was a 'colourd ray in the light before refraction.' Indeed, Newton did not seem consistent in his account of what this trial showed. In February, he said that it demonstrated that there were differently refrangible rays in light without reference to colour; in June, when publicly answering Hooke, he said that it demonstrated that 'rays of divers colours considered apart do at equall incidences suffer unequall refractions,' so raising the issue of specific colour. (Schaffer 1989: 86)

While Bazerman views the controversy as an occasion for Newton to develop persuasive stylistic means to communicate his factual claims, Schaffer indicates that the nature of those claims was itself at issue. In significant respects, Newton respecified the actual content of his assertions in response to the rhetorical exigencies of the dispute. Newton was engaged in articulating a new model of the relations between experimental fact and theoretical interpretation at the same time that he was seeking to have his claims accepted as facts. In this situation, rhetorical decisions evidently embraced more than just questions of style. Designations of what were factual claims and construals of their content were themselves constructed with available rhetorical resources and were subjects of negotiation in the ongoing debate.

This dimension of the rhetorical construction of facts emerges most clearly when we scrutinize episodes of prolonged controversy. A similar case, in which the issue of the rhetorical form of discourse was bound up with that of the factual claims, can be found in the debate surrounding the "chemical revolution" of Antoine Lavoisier at the end of the eighteenth century. On one reading, Lavoisier's radical challenge to the chemistry of his time appears to be embodied in a series of innovative factual claims, concerning the nonexistence of phlogiston (the traditionally identified principle of inflammability), the reinterpretation of combustion as a process of oxidation, the designation of water as a compound, and so on. But, as a complementary part of their program, Lavoisier and his allies also advanced a new nomenclature for chemical substances and a new model of chemical discourse as a "demonstrative" process of reasoning akin to geometry (Golinski 1992b, 1994). In the subsequent lengthy controversy, linguistic usage and rhetorical form were themselves called into question. Lavoisier's British opponents, such as Joseph Priestley and James Keir, challenged his style of writing at the same time as they cast doubt upon his empirical claims. Lavoisier's discourse embedded designations of certain statements as "facts" that the

British chemists could see only as "hypotheses" or (questionable) interpretations of experimental results. Hence, Keir wrote to Priestley that he wished the French chemists "would relate their facts in plain prose, that all men might understand them, and reserve the poetry of the new nomenclature for their theoretical commentaries on the facts" (quoted in Golinski 1992b: 238).

For Lavoisier, language was to serve as an "instrument," analogous to the material apparatus of the laboratory, to be used in the construction of a demonstrative discourse. Lavoisier devised a language that would serve his aims of securing a higher standard of proof than had previously been customary in chemistry. As scholars such as Wilda Anderson (1984), Trevor Levere (1990), W. R. Albury (1972), and Lissa Roberts (1992, 1993) have shown, he resorted to the philosophy of Etienne Bonnot de Condillac to validate this position. Condillac had proposed the vision of an "analytical" language in which signs corresponding to simple ideas would be manipulated to yield exhaustive knowledge of a subject. Lavoisier built upon this a nomenclature designed to discipline its users so that they would be constrained to accept his own account of chemical composition. He proposed that "A well-formed language, a language in which one will have captured the successive and natural order of ideas, will bring about a necessary and even prompt revolution in the manner of teaching. . . . [Chemists] will have either to reject the nomenclature, or else to follow irresistibly the route that it will have marked out" (quoted in Anderson 1984: 177).

Lavoisier's British opponents, particularly Priestley and Keir, resisted the new nomenclature as an attempt to usurp the common language of what they envisioned as an egalitarian scientific community. For Priestley, this community was bound together by use of the currency of descriptive factual discourse. To impose a language shaped by a particular "system" would be to disrupt the bonds of trust, so that speakers and writers would no longer be able to repose credit in one another's reports. To Keir, similarly, the new nomenclature served the interests of a "sect" seeking to advance its own "peculiar opinions"; it could not be considered "as the general language of chemistry." To him, what seemed illegitimate about the new terminology was precisely what Lavoisier and his colleagues had considered its strength – namely, that "we cannot speak the language of the new Nomenclature, without thinking as its authors do" (quoted in Golinski 1992b: 246).

The chemical nomenclature of Lavoisier and his allies was, eventually, accepted in Britain as elsewhere. With some modifications, it was gradually taken up by the majority of British chemists, although many of them pointedly distinguished the question of the value of the terminology from that of the validity of the "facts" Lavoisier had reported. A study of the resistance the nomenclature originally encountered has, nonetheless, some value. It helps us to discern how issues of rhetorical form came to be connected with questions of fact, and how the connection could be challenged. Both the specific linkage of style and content that Lavoisier put forward, and their uncoupling by his critics, were historical accomplishments, which should be studied as such rather than taken for granted. Even British chemists who came to accept what they took to be his basic factual claims, such as William Nicholson and William Higgins, reflected that both sides of the debate could apparently frame interpretations of experimental facts in their own theoretical language. Nicholson drew the conclusion that chemists could not be converted to the new system by "the direct force of right reasoning," but would have to be won over by a more indirect persuasive strategy. For these chemists, agreement with Lavoisier's empirical assertions did not imply a concomitant agreement that scientific language ought to be used in the way he had proposed (Golinski 1992b: 247–250).

Consideration of this kind of episode can also convey a theoretical point: that rhetorical purpose does not correspond to reception or determine how texts are read. Lavoisier's aim of using language as a particular kind of instrument of persuasion was resented by his critics. The general implication here is that rhetorical analysis of scientific texts can be significantly aided by investigating the ways in which they were actually read. Although we should not expect readers' responses to correspond directly to the rhetorical aims of the author (as the incidents of controversy show), focusing on the former offers a way to discriminate among rhetorical analyses that might otherwise be elaborated in an uncontrolled manner. The links that are exposed in the course of controversies – between factual claims and stipulations of method, or between discursive practice and other kinds of action, for example – can help substantially in the task of discerning what the rhetorical aim of a writer or speaker actually *was*. This is to say that rhetorical analysis should be framed within a more comprehensive process of *hermeneutics*, the act of interpretation being directed both at the author and the readers of texts. In this way, analysis of the rhetorical construction of scientific discourse, using the categories of form, audience, and situation, can be complemented by investigation of the processes of interpretation undertaken by actual audiences in specific contexts. How this might be done is the subject of the next section.

STEPPING INTO THE CIRCLE

In the light of the previous discussion, we might say that all language use is rhetorical, in the sense that one of its functions is to attempt to persuade. Even the baldest factual utterance has the implicit aim of convincing some putative listener or reader that what it asserts is true. This is not to say that there are not many different kinds of persuasion – from holding a gun to someone's head to laying out a reasonable argument – but that the task of persuasion is always part of the activity of communication. Be that as it may, however, it is arguable that to look at language as *only* rhetorical in function is to miss something else of vital importance. Aside from securing assent, language also works in other ways; crucially, it conveys meaning.

This signifying function of language can more plausibly be regarded as a complement to its rhetorical role than as a polar opposite. Neither function excludes the other, but the two point toward distinct modes of analysis of discourse and its context. While rhetoric focuses on the purposes of speakers and writers – their appropriation of available resources to advance specific claims in particular contexts – hermeneutics is concerned with how meaning is constructed by the interpreters of discourse. History is an inherently hermeneutical enterprise in two senses. First, historians are concerned with recovering the meanings that past discourse had for those who originally produced and understood it. Second, they are themselves participating in an interpretive undertaking, in which their own contexts shape their specific construals of the past. Historians professionally circumnavigate the "hermeneutic circle" in which the meaning of past discourse is revealed by understanding its context, which in turn is known by interpreting the historical texts in which it is represented. Engagement in hermeneutics frequently raises issues of "reflexivity," in which writers shift between advancing their own interpretations and commenting upon their own situation as interpreters. These issues seem to arise inevitably once one recognizes that historians are always situated inside the domain in which interpretation is accomplished.

Writers within the tradition of philosophical hermeneutics have reflected upon this process. Hans-Georg Gadamer, who relies heavily upon the philosophy of Martin Heidegger, has challenged the "dogmatic objectivism" that envisions historians being able to detach themselves from their own contexts to gain a more accurate understanding of the past (1976: 28). Being rooted in a certain linguistic tradition is part of the human condition, in Gadamer's view, and is not to be evaded by facile gestures at objectivity. Paul Ricoeur, on the other hand, has identified a "positive notion of distanciation" that opens up the possibility for the historian to gain at least some distance from his or her own standpoint while engaged in interpretation of the past (1981: 131–144). Although it seems to give more philosophical encouragement for their enterprise, Ricoeur's position offers no more specific guidance than Gadamer's as to how historians might improve their methods.

Nonetheless, against the background of these philosophical discussions, the vision of history as a hermeneutical enterprise has recently been gaining ground. This is partly a response to the increased popularity of anthropological methods in cultural history. Keith Michael Baker (1982) and William J. Bouwsma (1981) have independently called for intellectual history also to be subsumed within a "history of meaning," which would embrace all aspects of culture and all forms of symbolic expression of human experience. Baker explains, "The intellectual historian analyzing a text, concept, or movement of ideas, has the same problem as the historian faced with any other historical phenomenon, namely to reconstitute the context (or, more usually, the plurality of contexts) in which that phenomenon takes on meaning as human action" (1982: 197–198).

Understood in this sense, the "history of meaning" connects with the emphasis on localization that is characteristic of constructivist analyses of science. The link can be made via Wittgenstein's notion of "language games," in which the meanings of words or phrases are found by studying their uses in localized "forms of life." Thus, the historian works to connect particular local understandings of the elements of scientific discourse with the interests and practices of specific communities of language users. Rhetorical analysis, with its focus on how language is manipulated by authors and speakers, cannot fully address this issue. What is required is exploration of the variety of interpretations of a particular unit of discourse, the range of readings given to a specific text, or the spectrum of appropriations of a certain image. As these different local interpretations are distinguished, the topic of translations between the separate linguistic domains also emerges as a problem worthy of investigation. A fine example has been given in Andrew Warwick's studies of the readings given by mathematical and experimental physicists at Cambridge to the 1905 paper in which Albert Einstein enunciated the special theory of relativity (Warwick 1992, 1993). Warwick has shown how acute the problems of translation were here, and how the key to grasping the different interpretations of the text is an understanding of local cultures of theoretical and experimental work. The paper later hailed as announcing a revolutionary new theory of physics was simply not read that way by the British physicists, who did not recognize its supposedly innovative features and either dismissed it from consideration or assimilated it to their own programs of research.

To frame history of science as a hermeneutical activity in this sense, however, raises the problem of identifying the relevant contexts. Talk of Wittgensteinian "forms of life" does not resolve the matter of what the pertinent settings for understanding scientific activity might be. In practice, all sorts of contexts have been invoked in historical or sociological analyses, from nation-states, to disciplines, to "core sets," to the coworkers in a single laboratory. Presumably, hermeneutic attention might be focused on any one of these levels, to discern the ways in which meaning is made in that community, how the group's linguistic boundaries are sustained, and how translation is managed across them. One situation in which hermeneutic analysis might be expected to be particularly productive is that in which different disciplines are carving out their territory in overlapping domains and are engaged in competition to determine the language in which certain phenomena should be described. In this situation, and in others, communication between disciplinary specialists and a nonexpert public might also be of crucial importance. In a seminal article on the hermeneutics of science, Gyorgy Markus has therefore suggested that the construction of meaning in communications between experts and their public audiences should be closely scrutinized. Since the early nineteenth century, Markus notes (1987: 19–29), specialist and professional scientists have been required to engage in some justification of their activities to audiences that ultimately fund their work but do not share their kind of expertise. The ways in which meaning is constructed in these interchanges shapes both the understanding of science among the lay public and the formulations of experts themselves. Even discourse within specialist communities might well be affected by this process.

The creation of meaning by translation between linguistic communities has frequently been discussed in terms of the workings of *metaphor*. The Aristotelian notion of metaphor as the carrying over of a word from its original application to a new object of reference is sometimes invoked to describe how scientific language may be taken up by lay communities. Terms such as "affinity," "hysteria," "evolution," "entropy," "relativity," and many others are said to have originally been coined with quite precise applications within technical discourses but to have subsequently been appropriated for more general – presumably less precise – usage among nonexperts. The problem with the model stated in these terms, however, is that it assumes the possibility of a precisely denotative language within an unproblematically bounded scientific community. For a number of reasons, this assumption has recently come to seem implausible. In the wake of the controversy studies, the boundaries of scientific communities and assessments of what is good scientific practice are found to be frequently contested. Studies of the production of phenomena by localized instrumentation and skills have also called into question the supposition that a "technical" term always denotes exactly the *same* phenomenon. And a number of historical analyses have shown how scientific practitioners themselves have borrowed terms from other areas of culture, bringing their inseparable connotations into supposedly technical discourse. M. Norton Wise has shown, for example, how British writers on mechanics and other areas of natural philosophy in the 1830s and 1840s made significant use of notions such as "work" and "waste" that originally derived from the discourse of political economy (Wise

1989a, 1989b, 1990). It is not the case, then, that the meaning of technical terms is precisely fixed by scientists and subsequently messed around by everyone else.

Roger Smith's meticulous study of the meanings of "inhibition" in the nineteenth and twentieth centuries has provided an excellent example of this kind of history (R. Smith 1992a, 1992b). Smith shows that the word "inhibition" was embedded in popular discourses describing character and conduct that continued to hold sway within specialist communities. Even in the technical writings of C. S. Sherrington and Ivan Pavlov, where "inhibition" acquired a relatively specific meaning, it was not possible to avoid invoking its more general usages. And this was frequently done quite deliberately when psychologists addressed lay audiences (R. Smith 1992b: 242). As Smith notes, "a specialized community may reconstruct language for a particular purpose, but it does not thereby empty language of its wider meaning" (1992a: 228).

For Smith, the relevant wider meanings were primarily social ones. "Inhibition" encoded notions about social order and how it should be maintained, which can be shown to have informed the more "technical" psychological uses of the term. As Smith explains:

> [T]here have been two major senses in which "inhibition" has helped constitute knowledge about order. In the first, the word referred to a hierarchical arrangement in which a higher power arrests or depresses a lower power. In the second, the word portrays how more or less equal powers compete for limited resources. The former was used in theories about organizational levels in the brain, the latter in psychologies describing the organization of consciousness or behavior. These usages indicated that the word denoted processes in the body or in the mind. At the same time, they indicated that the word connoted relations of power within societies: the arrest of lower by higher agents or the competition of the economic marketplace. (1992a: 12–13)

Smith's careful and thorough hermeneutical investigation succeeds in tracking the meanings of "inhibition" across the borders between scientific and other communities. He thereby convincingly demonstrates how "words carry social values into the heart of science" (1992a: 6). And he voices an important methodological proviso: He declines to distinguish literal (or denotative) from metaphorical (or connotative) meanings. No word, he suggests, could be so precisely and unambiguously attached to a single reference as to escape entirely from its connotative meanings: "It was not possible, in the last resort, to draw a firm distinction between what inhibition denoted and what it connoted. There was no ground independent of the web of metaphorical meanings" (237).

Smith's point here is that use of an overly rigid model of metaphor would suggest that the boundary between precise denotative usage by specialists and loose connotative usage by others is perfectly secure, whereas historical investigation uncovers a considerable degree of permeability in both directions. James J. Bono has made a coincident point, using some of the terminology of contemporary literary theory, specifically the "deconstruction" that takes its lead from the work of Jacques Derrida. Bono identifies the supposition that he calls "the role of metaphor," according to which "the special insight of scientists contains and controls the language and metaphors constituting scientific discourse" (Bono 1990: 60–61). Against this, he sets "the rule of metaphor," which gives "full weight to the dense reality of discourse and to the dissemination and proliferation of meanings within and beyond the boundaries of science itself." Bono explains his reasons for preferring the latter perspective:

> [C]omplex scientific texts and discourses constitute themselves through their intersection with other multiple discourses. Such intersection sets up interferences among various discourses leading to the dissemination of various meanings with the power to disrupt, to resist, and to transform the metaphors and deeply embedded tropological features of the languages of a given discourse. (1990: 61)

From this point of view, metaphors rule their users, rather than vice versa. Language is not under the control of scientists, or indeed of any authors. Meaning is constantly being disseminated and ambiguity cannot be altogether removed. Metaphorical translation is continually occurring, so that linguistic reference cannot be fixed with total precision or for very long. As Bono puts it, "Texts defy the efforts of their authors to control them in large measure because the tropological and rhetorical dimensions of language – which cannot be bracketed or stripped away – ensure a multiplicity of meanings and the possibility of continual reinterpretation" (1990: 66).

Although Robert M. Young has located himself in a Marxist tradition rather than a deconstructionist one, he takes a position similar to Bono's in relation to the topic discussed in his essay on "Darwin's metaphor" (1985). He there analyzes the uses of the term "selection" in the texts of Charles Darwin and in those of his critics and commentators. He insists that the "scientific" meanings of the word cannot be separated from its connotations in theological and philosophical debates, which were "constitutive, not contextual" in relation to evolutionary theory (Young 1985: 80). The argument of *The Origin of Species* moved from artificial selection, as practised by farmers and breeders, to the "natural selection" that Darwin proposed was responsible for the evolution of new species. In making the shift, Darwin quite deliberately retained the anthropomorphic and voluntaristic implications of the term "selection." Although he dropped the explicit image of a superhuman being, which had been present in his private notes, he continued to refer to a power of "Nature," "daily and hourly scrutinizing, . . . rejecting that which is bad, preserving and adding up all that is good; silently and insensibly working" (Darwin 1859/1968: 133).

The metaphor was a potent one, but it worked in ways Darwin does not appear to have anticipated. Some critics objected to the anthropomorphic connotations of "natural selection." Alfred Russel Wallace tried to persuade Darwin to drop the phrase and not to personify "Nature." "[P]eople will not understand that all such phrases are metaphors," he warned (quoted in Young 1985: 100). On the other hand, commentators such as Charles Lyell, John Herschel, and Asa Gray wanted to interpret evolution as a process guided by Providence; they therefore seized upon Darwin's anthropomorphic language as encouragement for their views. Although Darwin initially disclaimed the intention of advancing a providential theory, he became resigned to these theistic interpretations and professed himself happy that at least the uniformity of the natural laws governing evolution was being recognized. In later editions of the *Origin*, he defended his choice of metaphorical terms as necessary to communicate concisely with his readers and pointed to other scientific expressions (such as chemical "affinity" and gravitational "attraction"), which had originally had anthropomorphic connotations but subsequently lost them.

On Young's reading, then, Darwin himself had a sophisticated understanding of the importance of metaphors in scientific communication, and he was obliged to recognize that they sometimes conveyed meanings not intended by their author. Stepping back from Darwin's own point of view, Young is able to bring further elements into the picture. He shows how Darwin's use of anthropomorphic language in connection with natural selection reflected the continuing influence of the natural theology tradition, represented preeminently by William Paley. This was the tradition to which the more theistic of Darwin's professed adherents still clung. On the other hand, nonprovidential conceptions of natural selection were also quite plausibly derived from his works. The expression therefore welded together a quite diverse group of interpreters of Darwin's texts, some of whom explicitly retained a vision of the natural world as the product of divine design.

A deft contextual analysis such as this indicates the inadequacies of a simplistic notion of scientific metaphor. Metaphor is not something exterior to science, the result of lay people's misunderstandings of precise technical terms. Rather, it is a label for the creative function of translation among different linguistic communities, or for the creation of new meanings over time within a single community. Scientists continually recruit linguistic elements from other realms, and they cannot control the uses to which their own language will be put. In science, as in all discursive activities, metaphor is ineradicable; on the contrary, it is necessary for the creation of meaningful discourse. As Gillian Beer has put it, in justification of her own extended and very subtle readings of Darwin's metaphors: "Space, expansion, forecast; these are the powers offered by metaphor, whether scientific or literary – and they are powers as important as the correspondence, similitude, and exactness of measure, which we habitually look for" (Beer 1983: 92).

Beer has shown how much can be achieved by tracking Darwin's borrowings and lendings at the level of language. She has deployed a sensitive literary imagination and a very wide knowledge of Victorian literature to trace the impact of Darwin's reading on his imagery and his choices of figures of speech (Beer 1985). She suggests, for example, that Darwin's vocabulary for discussing the relations between artificial and natural selection might have been conditioned by his reading of Renaissance treatments of art and nature in such authors as Montaigne and Thomas Browne. Similarly, his somewhat refractory interpretation of Thomas Malthus, in which the dire threat of overpopulation and famine was converted into the motor of evolution, might have been indebted to John Milton's vision of fertility as the source of the richness of the natural world (though Paley would be a more obvious and immediate source for such a vision). Turning from Darwin's reading to the influence of his writing, Beer has traced the diffusion of the language, imagery, and narrative techniques he inspired into the works of Victorian novelists (1983). She practices a complex hermeneutics, in which the qualifications and ambivalences characteristic of deconstruction are prominent. The aim is to go beyond general statements about the influence of ideas to grasp the role of such linguistic elements as analogies and images, narrative devices and descriptive mechanisms, plot structures and characteristic lacunae. Although Beer acknowledges the possibility of some degree of control of meaning within specialist communities, her focus is on "the excluded or left-over significations of words [which] remain potential and can be brought to the surface and put to use by those outside the professional agreement as well as by those future readers for whom new historical sequences have intervened" (Beer 1985: 544).

Beer's method is largely irreducible to summary. Her claims are mostly advanced suggestively, with an awareness that they have not been (and perhaps cannot be) demonstrated as conclusively as historians might wish. Rather than attempting to condense her findings here, I shall offer, in the remainder of this chapter, three possible keys to hermeneutic analysis of scientific discourse. Each one highlights a particular element of scientific language that can be tracked in its different meanings and usages. The purpose of the typology is to move beyond somewhat generalized discussions of metaphor, to suggest how scrutiny can be directed to more specific features of discourse, and to provide examples of how this has been done. (The typology develops and modifies one I have articulated elsewhere: cf. Golinski 1990b.) The three categories of analysis are: first, *semantic*, in which the meanings of certain key words in particular local contexts are discerned; second, *semiotic*, which is concerned with interpretations of symbolism and imagery; and third, *narratological*, which focuses on the assemblage of linguistic elements into narrative structures or stories. There is no suggestion that these three categories exhaust the possible ways in which linguistic usage can be analyzed in historical hermeneutics; they are simply advanced as convenient ways of recognizing the level on which linguistic meaning is being located.

Semantic analysis can be exemplified by Mi Gyung Kim's treatment of the meanings of the terms "atom" and "molecule" in nineteenth-century organic chemistry (1992a, 1992b). Kim considers a situation in which specialist communities were quite well established, and where one might expect the meanings of technical terms to be fixed and common to different national languages. However, her analysis discloses important local variations in meaning, which harked back to different national traditions of chemistry that remained pertinent well into the nineteenth century. Thus, French chemists at the beginning of the century had tended to use *atome* and *molécule* synonymously; but they preserved a distinction between the two terms when translating John Dalton's "atom" because they recognized its relation to his particular assumptions and practices of stoichiometry (the chemistry of combining proportions). It was only later that the meanings of the two words "became relatively stabilized within organic chemistry, which designated a molecule as a group of atoms" (Kim 1992b: 399). Amedeo Avogadro used the French term *molécule* in his classic paper of 1811, in which he proposed that equal volumes of gases under the same conditions would contain equal numbers of particles. But this usage, Kim argues, was at variance with leading French chemists' assimilation of Dalton's "atom," with which Avogadro's *molécule* was effectively synonymous. Kim suggests that, "Avogadro's true-to-dictionary, yet idiosyncratic use of 'molécule' betrays his isolation in the periphery of the French intellectual empire. He was well-informed on the contemporary English and French chemical literature, but had no personal contact with the leading French chemists" (406).

Kim draws the conclusion that, "The tortuous path of chemical atomism in the nineteenth century demands that historians pay close attention to the languages of historical actors and intercultural transmissions: words are tricky mediators of thought and action" (1992b: 427). Her historical semantics displays an admirable sensitivity to the subtle nuances that differentiate linguistic usages between national languages and between distinct research agendas (such as crystallography and stoichiometry). The different subdisciplines preserved distinct ontologies and programs, even while they shared terminology. Kim adopts Peter

Galison's notion of a "trading zone" to characterize the process by which linguistic elements were exchanged across boundaries that separated discrete realms of meaning. Galison's term – itself a metaphor borrowed from anthropology – designates a place where signifying tokens are passed between communities which may have quite different understandings of the meaning of what is traded, but both of which value the process of exchange (Kim 1992b: 428; Galison 1989).

We can compare Kim's semantic analysis of the different meanings of terms in separate specialist communities with Roger Cooter's remarkable study of phrenology in nineteenth-century Britain (1984). Cooter focuses on a more multivalent discourse, dispersed among largely nonspecialist communities at various levels of society. He adopts what we can call a semiotic approach, aiming to discern the symbolic meanings of phrenological concepts and images for those who used them. As with Smith's analysis of inhibition, phrenology is viewed at least in part as a symbolic projection of social perceptions and ideals into the realm of mental science, although it is not supposed to have been a straightforward or unambiguous reflection of social reality. On the contrary, Cooter stresses that phrenology was a highly flexible semiotic resource, its ambiguities being visible ever since its origins in the 1790s in the works of F. J. Gall, who attracted followers from among both radical materialists and anti-Enlightenment mystics (1984: 39–41).

By the 1840s, on Cooter's reading, defenders of the subject who craved bourgeois respectability had succeeded in distinguishing "pseudo-phrenology" from the genuine article and largely deflected charges of atheistic materialism onto the former. The works of George Combe provided symbolic resources for members of the working class seeking self-improvement, who were keen to embrace a system that recognized differentiated mental abilities but did not identify them with inherited positions in society. To explore how Combe's writings appealed to this readership, Cooter engages in a "decoding [of] the science's signs and symbols," which reveals its operation as a "multifaceted symbolic resource" (1984: 110, 119). The phrenological model of a hierarchy of mental faculties gave priority to reason over the emotions and encouraged those who sought to gain a higher social status by virtue of their intellectual skills. Phrenologists such as Combe proposed that social standing should be determined by one's mental abilities, and that those abilities could be improved by discipline and effort. These lessons could be read from texts, from lectures, and even from the widely distributed images of the phrenological head, with its surface divided into areas corresponding to the locations of the different mental faculties.

Combe's version of phrenology thus appealed to the middle-class managers of the Mechanics' Institutes where it was taught; they were aware of its entertainment value but also saw it as promoting discipline

and self-improvement among the working classes. To working-class au-
todidacts themselves, the subject promised a comprehensive philosophy,
relatively easily mastered but supposedly yielding profound insights
into human life. In practical terms, phrenologists offered advice about
how to decide upon a career, how to find a suitable marriage partner,
or how to choose a reliable servant. Cooter notes, however, that, as a
"public meaning system" or a "secular religion," phrenology offered a
set of symbols for representing society and social aspirations that also
constrained the ways in which those aspirations could be expressed:

> These symbols could be consented to readily since, quintessentially, they
> were personal, intimate, common, flexible, and "real" (because of their
> external palpably empirical nature). But in so accepting, one consented to
> interpret experience in certain ways; one consented, if not to wholly en-
> dorse certain "objectified" structures and commit oneself fully or in part
> to a certain "natural" orientation to action, then at least to allow certain
> other orientations to be obscured. (Cooter 1984: 190)

The potency of phrenology as a symbolic system was rooted, Cooter
argues, in its appearance as a representation of natural facts. It was possible for it to be "taken up noncritically, unreflectively, as factual knowledge of nature" (1984: 191). Although historical analysis can exhibit its
notions and images as symbols of social order and individual aspiration,
to those who accepted them they were "facts" and therefore "impossible
to frame for debate" (192). It is in this connection that Cooter reaches for
Antonio Gramsci's concept of "hegemony" to capture the sense of a discursive system that was willingly embraced by its followers but insinu-
ated a hierarchical vision of society under the apparently unquestionable
guise of natural facts.

Cooter is remarkably subtle in teasing out the ambiguities of phrenological symbolism and suggesting its meanings to different social actors.
His analysis cannot be accused of any kind of crude reductionism in its
application of the concept of hegemony. It is, however, frequently – and
perhaps necessarily – conjectural. Although a great deal of empirical evidence for phrenological activity is adduced, the analysis often goes beyond this when it suggests its meaning. Insofar as he argues that
phrenology constituted a "mystified mediation of ideology," Cooter is
engaged in proposing a meaning for phrenology that its devotees did
not recognize – that, according to his analysis, phrenological discourse
prevented them from recognizing (1984: 192). The metaphors for social
order and hierarchy that are encoded within phrenological symbolism
can only be perceived as such from the vantage point of the analyst who
is equipped to decode facts as "naturalized" social relations; the historical subjects themselves cannot be expected to voice any such realization.

Cooter is therefore doing something more than locating different interpretations of phrenological discourse in specific social groups; he is proposing that the discourse comprises some representation (albeit an inadequate one) of social reality. The model of socioeconomic "base" and cultural "superstructure," with which the concept of ideology has long been associated in Marxist theory, is in the background here. The relations between the two layers are, in Cooter's analysis, highly mediated ones, but they nonetheless configure phrenology's ideological role. Cooter notes that he is not saying that working-class learners of the subject were forced to accept "the naked consciousness of the dominant class," but that they did consent to "the ideological configurations and priorities dependent upon and implicit to the dominant material relations" (1984: 192). Insofar as he asserts the possibility of this kind of decoding of the ideological representation of social reality, Cooter's hermeneutics goes beyond situating meanings in historical communities and lays claim to the historian's ability to determine what the *real* meaning of phrenology was. This is a strategy of interpretation that places the historian at the heart of the hermeneutic circle and depends for its legitimacy upon acceptance of the historian's own theoretical presuppositions.

Greg Myers takes an apparently more neutral stance in exemplifying what I shall call narratological hermeneutics. Here, the analytical focus is upon the meaning that is created by assembling linguistic elements into coherent narratives. Myers discusses the narrative strategies employed in E. O. Wilson's *Sociobiology* (1975), with its contentious proposals for understanding human behavior by comparison with that of animals, and in the writings of Wilson's critics (Myers 1990: 193–246). Following the lead of the controversy studies, Myers displays how each side of the dispute deconstructed the claims of the other. While Wilson proposed that studies of animal behavior could throw significant light on human conduct, his critics accused him of attempting to justify social inequality on the basis of spurious analogies with the animal world. Myers finds both sides in the debate accusing the other of quoting selectively and out of context. He suggests that this is inevitable, because "there is no context large enough to guarantee that a statement will have just one meaning, the intended meaning, that it will speak for itself" (1990: 219). The controversy thus took the form of a struggle by each side to situate the statements of the other in a convincing narrative context. Wilson portrayed his critics as political ideologues who could not sustain an argument on the level of scientific evidence; they located him in a tradition of conservative thinkers who "naturalize" prevailing social arrangements by identifying them with instinctive animal behavior.

A controversy such as this renders texts "opaque," Myers proposes, and thereby illuminates the processes by which they more typically function as transparent bearers of facts. Seeing the statements of Wilson and others wrenched from their original settings and inserted into alternative narrative contexts, our attention is directed to the means by which they were originally contextualized by their authors. In this connection, Myers dissects the narrative form typical of natural history. He points out how Wilson's ethological passages usually describe episodes of encounters between an observer and individualized members of a certain animal species, usually with some circumstantial details and some note of the reaction of the observer. Wilson skillfully assembles these descriptions together into evolutionary narratives – stories of how certain behavior patterns arose through the course of evolution. In these second-order stories, the subjects of the action are not observers or individual animals, but abstract entities, such as species struggling to survive or (Darwin's metaphor again) "nature" as a superhuman selector. Thus, according to Myers, Wilson makes "his model seem to correspond with perceived reality . . . by inserting a narrative of natural history, which we associate with reality, within a narrative of evolution, which we associate with model building" (1990: 195).

In the course of the controversy, this narrative strategy was deconstructed by Wilson's critics. Elements of his text were relocated within new interpretive contexts, so that their meaning was changed. They were presented, not as descriptive passages of natural history that accumulate to confirm evolutionary hypotheses, but as just one more attempt to convince people that inequitable social arrangements are "natural." Wilson's tales of evolution were resituated as parts of a story of oppression and its naturalistic justification. It became impossible thereafter to settle on a consensual interpretation of what Wilson's text actually means. Consensus was prevented, in part, because each side found it profitable to caricature the other's position as tainted by emotions or political interests. Neither side, however, could admit a political aim of their own, as that would amount to surrendering one's credentials in an argument that was ostensibly only about scientific facts.

Myers displays the role of competing narrative constructions in the controversy by declining to choose between the two sides. Adherence to Bloor's symmetry postulate leads him to refuse to assign political or social motives to just one side in the dispute. Hence, he does not characterize Wilson's sociobiology as a naturalization of elitist or racist social attitudes; nor does he echo Wilson's view of his opponents as scientifically incompetent and motivated by a radical political agenda. He rejects, in other words, the break with neutrality that would be involved in ascribing an ideological function to Wilson's sociobiology. Instead, the focus is on the process of construction by which elements of language are put together into meaningful stories by authors on both sides of the debate. The artificiality of the narratives thereby created is exposed by this analysis, but the role (if any) that social interests play in the process remains obscure.

In comparison with Cooter's, Myers's analytical stance is declaredly a more "symmetrical" one, and it is certainly illuminating of at least certain features of the controversy he discusses. It may well be advisable to set aside issues of what social motivations were operative when analyzing a debate in which that question is precisely at issue between the disputants. Myers mounts a credible defense of his stance in refusing, for example, to debunk Wilson's sociobiology as merely disguised ideology (1990: 247–259). This does not, however, amount to adopting a position of absolute objectivity or taking the historian out of the hermeneutic circle. As Myers is well aware, he himself is engaged in constructing meaning by assembling linguistic fragments into his own narrative structures. All historical hermeneutics, whether its focus of analysis is semantic, semiotic, or narratological, is obliged to make some acknowledgment of the problems of reflexivity that are consequent upon the historian's own position within the hermeneutic circle.

5

Interventions and Representations

The operations and measurements that a scientist undertakes in the laboratory are not "the given" of experience but rather "the collected with
difficulty." They are not what the scientist sees – at least not before his
research is well advanced and his attention focused. . . . Science does not
deal in all possible laboratory manipulations. Instead, it selects those relevant to the juxtaposition of a paradigm with the immediate experience
that that paradigm has partially determined.
Thomas Kuhn, *The Structure of Scientific Revolutions* (1962/1970: 126)

INSTRUMENTS AND OBJECTS

As I noted at the beginning of the previous chapter, a comprehensive
study of scientific activity cannot confine itself to the level of discourse.
Scientists talk and write a great deal, but that is not all they do. In this
chapter, I turn the spotlight onto their other practices: the manipulations
of materials and apparatus that constitute the work of experimentation,
and the nondiscursive means of representation they employ, particularly
visual images in their many forms. Constructivists have had significant
things to say about the topic of material practice in experimental science.
They have considered the relations between the tools of investigation
and the objects at which experimental research is directed, and the curious ways in which tools and objects can merge, become distinct, or
change places. They have also tried to understand the use of visual representations as itself a kind of practice, an activity continuous with the
other activities in which researchers engage. Scrutiny of visual and material practices has shed further light on "the problem of construction"
– on the ways in which knowledge constructed with specific localized
resources is made reproducible at other sites.

Since the seventeenth century, when natural knowledge was first systematically pursued through experiments, scientific research has been
conducted in a purpose-built world of specially manufactured instruments. By instruments are meant the material tools the human investigator uses to disclose, probe, isolate, measure, represent, or otherwise
bring to attention the objects of investigation. Experimental phenomena

are produced and reproduced only by specific assemblies of apparatus; some philosophers have talked of them being literally "made real" by the complexes of instrumentation within which they are constituted (Hacking 1983: chap. 13; Ihde 1991). Gaston Bachelard's term "phenomeno-technique" captures the sense in which "natural" phenomena and experimental hardware are produced and reproduced together; although, as John Schuster and Graeme Watchirs (1990: 21–25) have pointed out, Bachelard did not accord to the experimenters the degree of flexibility in interpreting their results that constructivist sociology has emphasized. This emphasis has brought into the picture the resources of specific local settings, such as laboratories, in which phenomena and hardware are reproduced and interpreted.

Practitioners of the sciences work hard to sustain a distinction between their instruments and the objects they bring to light. They try to establish that the phenomena elicited are not mere artifacts of a particular apparatus and can be reproduced elsewhere with other and preferably quite different tools. Any serious analysis of experimental practice has to recognize the labor devoted to maintaining the distinction between objects and instruments. But the distinction can be shown to be a constructed – not an absolute – one. Instruments generally start out as objects of investigation before they are felt to be understood and can be trusted as means to produce new phenomena. When they become taken for granted as tools, they can then be employed together with other instruments in complex systems that configure objects so as to make them available for observation and manipulation. These systems can assume an importance extending well beyond the site of original research; they form crucial components of the infrastructure that makes possible reproduction of experimental phenomena away from their places of origin. It is always possible, however, for the distinction between object and instrument to break down and for doubts to arise as to the reliability of the apparatus, which then reverts to the status of an object of investigation, its phenomena at least temporarily suspected of being artifacts.

A number of recent historical studies have been devoted to the creation of instruments, illuminating the ways in which they are made meaningful in particular social and cultural contexts (Gooding, Pinch, and Schaffer 1989; Mendelsohn 1992). Since Francis Bacon called for the use of "instruments and helps" for the hand and the mind, experimental science has pursued natural knowledge by manual manipulation, creating a technology to serve its cognitive ends. In the seventeenth century, many new devices were put to this use (Hackmann 1989; Bennett 1989; Van Helden 1983). Measuring and calculating tools were adopted from the mathematical arts and applied to the purposes of natural philosophy, most spectacularly in astronomy, where accurate measurements of the positions of the planets came to have profound cosmological significance.

Optical instruments also came to the fore with the inventions of the telescope and microscope. Robert Hooke was referring primarily to optical discoveries when he lauded the correction of the weaknesses of the senses by instruments, or "the adding of *artificial Organs* to the *natural*" (quoted in Vickers 1987: 102). In addition, a whole new class of apparatus was specifically manufactured for the investigation of natural phenomena. What came to be called "instruments of natural philosophy" included new measuring devices, such as the barometer and thermometer, and machines that created previously unknown effects, such as the air pump and the electrical generator. As J. A. Bennett (1986) and Mario Biagioli (1989) have argued, these devices were increasingly seen as having epistemological value as the practitioners of the traditional mathematical sciences, such as navigation and architecture, earned higher social status. Natural philosophers found themselves in relationships of patronage of – or partnership with – the makers of these instruments. Recognition of the value of material artifacts for making knowledge brought with it a series of problems of labor organization, as well as the practical difficulties of reproducing the apparatus reliably.

Initially, each innovation was likely to be contested. Shapin and Schaffer (1985) have shown this by tracing Hobbes's objections to Boyle's air pump. For Hobbes, a device of this kind was simply not an appropriate means of producing philosophical knowledge: "not every one that brings from beyond seas a new gin, or other jaunty device, is therefore a philosopher. For if you reckon that way, not only apothecaries and gardeners, but many other sorts of workmen, will put in for, and get the prize" (Shapin and Schaffer 1985: 128). Notwithstanding these objections, the authors also show how Boyle's pump became accepted and was used to communicate knowledge of the phenomenon of air pressure beyond the sites of the original experiments in Oxford and London. Their account emphasizes the difficulties encountered in the project of replication, drawing attention to the tacit knowledge that needed to be transferred to enable machines to work properly at other locations. But they also exhibit the part played by the materiality of the apparatus itself, as replicas were shipped from place to place. Evidently, given the right circumstances and human support, the pump could display a degree of "agency" (to use Andrew Pickering's term) in making reproducible the phenomenon of air pressure.

Albert Van Helden (1994) has told a formally similar story about the events surrounding Galileo's somewhat earlier innovation of the astronomical telescope. It has long been known that the first telescopic observations of the heavens, especially of the moons of Jupiter, were disputed. Van Helden has documented the steps Galileo took in the 1610s to overcome this opposition. On one occasion, he visited a group of skeptics in Bologna and tried to teach them to observe with his instrument – with mixed results. He circulated copies of his telescope to other investigators, though his strategy was primarily to send (somewhat inferior) devices to prospective patrons, rather than to allow potential rivals (such as Kepler) to get their hands on them. Galileo did, however, have some success in winning over Jesuit astronomers and members of the Accademia dei Lincei in Rome, to whom he demonstrated his discoveries in person.

Although much work remains to be done, historical research such as this is beginning to uncover the material culture within which the new experimental philosophy was constructed in the early-modern period. A range of specially fabricated apparatus was developed to extend the reach of the senses or to create novel physical conditions, and this technology was refined considerably and distributed quite widely among investigators and patrons. The material culture was sustained by particular social conditions: patrons who supported the manufacture of devices as "curiosities of art," natural philosophers willing to cross the line that had previously divided them from practitioners of the mechanical arts, and artisans who acquired the refined skills demanded by the new market. The seventeenth century saw the origins of a long-lasting relationship between the practitioners of experimental science and the manufacturers of the instruments their work required (Warner 1990). These restructured social relations provided the conditions in which the new apparatus could be made, circulated, and put to use.

In the eighteenth century, the dependence of experimental philosophy on specific apparatus deepened, and the social links between researchers and those who made scientific instruments were strengthened. Certain fields of investigation, such as electricity, relied upon the use of specialized devices for opening up new phenomenological domains. In others, such as chemistry, advances in apparatus also played a crucial transformative role. Makers of instruments frequently took the lead in exploring the commercial market for experimental science, pioneering the novelty of popular demonstration lectures and manufacturing such devices as orreries and barometers for use in middle-class homes. Commenting on the use of the orrery (a mechanical model of the solar system) in 1713, the journalist Richard Steele echoed Hooke's language of the 1660s: "It is like the receiving a new Sense, to Admit into one's Imagination all that this Invention presents to it with so much Quickness and Ease" (quoted in Schaffer 1994a: 159). In the century of the Enlightenment, natural philosophers and instrument makers were partners in bringing experimental science into the emerging public sphere (Porter et al. 1985).

The following century saw a radical alteration in the material culture of science, so that the practice of experimentation became even more closely bound to a much more complex technology, a fact that points up the importance of the historical transformation sometimes identified as the "second scientific revolution." In all the fields of physics – mechanics, thermodynamics, electromagnetism, and optics – and in the sciences of chemistry and physiology, new classes of instrumentation were created and developed, made possible by the new manufacturing techniques emerging from the Industrial Revolution. James Clerk Maxwell, reviewing the types of scientific apparatus available in 1876, analyzed them in the very terms used to describe contemporary industrial machinery. In each division of physics, apparatus was devoted to providing a source of energy, to communicating, storing, or regulating it, or to measuring its effects upon a defined system (Galison 1987: 24–27). In Maxwell's vision, physical science had become a form of industrial manufacture, dependent on the same mechanisms for energy conversion that characterized productive industry.

The vision of science modeled on industrial production embraced also the human workers in the nineteenth-century laboratory. As we saw in Chapter 2, the early part of the century saw the application of new types of apparatus to forge new disciplinary programs of research and student training. Precision measurement occupied a crucial place in the increasingly overlapping realms of training and research; it provided a means of inducting new practitioners into the disciplined work practices required of them, and it opened up extensive possibilities for investigation in the physical sciences (Wise 1995; Gooday 1990). John Herschel noted in 1830 that the availability of instruments made it possible to pursue experimental science as the quest for ever more precise measurements: "We are obliged to have recourse to instrumental aids, that is, to contrivances which shall substitute for the vague impressions of sense the precise one of number, and reduce all measurement to counting" (quoted in Warner 1990: 88). By the 1880s, when William Thomson (later Lord Kelvin) enunciated his famous doctrine that one could only say one had knowledge of a phenomenon if one had a means of measuring it, the association between instrumentation and quantitative precision had become firmly entrenched in the practices of the physical sciences. As Crosbie Smith and Norton Wise have emphasized, the link was forged particularly strongly in connection with the growth of electrical telegraph networks, on which Thomson and others worked to advance the enterprise of British imperial and commercial expansion (Smith and Wise 1989; cf. Hunt 1991, 1994).

Thomson's laboratory in Glasgow in the 1850s and 1860s had been the place where precision experimental work was first directed at serving the needs of long-range telegraphy, culminating in the 1866 success of the transatlantic cable, which gained him a knighthood. Thomson's triumph established the commercial importance of laboratory investigations of induction phenomena in cables and precise measurements of resistance (Smith and Wise 1989: chap. 19). These priorities were subsequently adopted elsewhere, especially at the Cavendish laboratory in Cambridge, directed by Maxwell after 1871. In Germany, other connections between experimental research and industrial production were equally vital, for example, Ernst Abbe's relationship with the optical company Zeiss in Jena or Hermann Helmholtz's links with the telegraphy and electrical power supply concerns of Werner Siemens in Berlin (Lenoir 1994).

In each case, the manufacture of equipment for the physics laboratory required the dedicated services of increasingly specialized instrument makers. In Glasgow, Thomson struck up a partnership with James White, who made devices to his specifications and also marketed them commercially. At the Cavendish, Maxwell initially sent instructions for instruments to London, but found this led to imprecision in translating designs into practice. He set about transplanting the successful Scottish model, employing a skilled instrument maker, Robert Fulcher, in the laboratory in 1877. Four years later, the Cambridge Scientific Instrument Company was formed to manufacture commercially the apparatus required for advanced physical research. Notwithstanding the importance of the Glasgow model, however, Maxwell resisted the imputation that the Cavendish would become a mere "manufactory" and insisted that precision measurement of electrical standards should be seen as having a moral value and a connection with the mathematical theory of electromagnetism that made it a worthy occupation for Cambridge students (Schaffer 1992).

Maxwell's achievement invites comparison with the simultaneous success of Michael Foster, who established a laboratory for experimental physiology in Cambridge in 1870. In Foster's laboratory, training and research were also focused around a specific coupling of instrumentation and phenomenon. Instead of quantitative measurement, the aim was to record changes in the dissected organs of snails, frogs, and other animals, to enable investigators to highlight the muscular origins of the heartbeat (Geison 1978). Physiological changes in dissected organs were registered in the form of continuous curves traced on the revolving drum of the kymograph, a crucial innovation imported from Germany, which was simultaneously reshaping laboratory work at Harvard (Borell 1987). Again, the phenomeno-technique was situated in a local culture that accorded it meaning. Foster, like Maxwell, took advantage of the strengthening climate of reform at Cambridge – the "revolution of the dons" – which enhanced support for laboratory-based research on the German model, in place of the traditionally entrenched mathematics and natural history.

The general implication of these case studies is that instrumentation, while it links individual settings to widely distributed networks of production and circulation of artifacts, still has to be assimilated and interpreted in each local context. Maxwell and Foster borrowed standardized apparatus and procedures from elsewhere – Maxwell from Glasgow, Foster from University College, London, where he had been trained by William Sharpey – but each saw the need to adapt them to the specific culture of Cambridge in the 1870s. To be granted authority in each setting, instrumentation has to be construed in terms of the local culture. Constructivist studies of the material culture of science have thus been obliged to grapple with the issue of what is specific to local settings and what may readily be translated between sites. What does the hardware carry with it – what capability does machinery itself have to "discipline" its users and change their practices – and what needs to be provided in each setting to supplement the material apparatus and make it useful? These are among the questions that the study of instrumentation in science raises.

Constructivist history has been inspired by the sociological studies of late-twentieth-century laboratories, which, obviously, encompass a material practice quite different from that of previous centuries. At its simplest, this is a matter of size. Galison indicates vividly just how *big* contemporary "big science" can be: "Through the 1930s . . . , most experimental work could be undertaken in rooms of a few hundred square feet, with moderate, furniture-sized equipment. But at Fermilab, . . . a herd of buffalo actually grazes in the thousand-odd acres surrounded by the main experimental ring, and individual detectors can cost millions or tens of millions of dollars" (Galison 1987: 14).

More significantly, late-twentieth-century experimental science has been characterized by an increasing complexity of its component elements. These are now connected with one another and with their context in much more heterogeneous ensembles. It is now routine for the instruments and techniques of many different sciences to be linked together as components of an "experimental system," often with differently skilled practitioners tending to their own parts of the ensemble. Thus, molecular biology requires techniques of bacterial and viral genetics, electron microscopy, biochemistry, developmental biology, and other fields. High-energy physics combines the resources of theoretical particle physics, electrical engineering, electronics, software design, and even personnel management. Referring to this striking heterogeneity of skills and material resources, which immediately confronts the analyst of contemporary experimental science, Adele Clarke and Joan Fujimura have called for the development of an "ecology of the contents of scientific knowledge . . . [and] of the conditions of its production" (Clarke and Fujimura 1992: 4). The comparison with ecology captures the aspiration to find hidden ordering principles behind the apparent confusion of heterogeneous things and activities in the contemporary laboratory. It also indicates the importance of relations of mutual interdependence between the different elements, like the connections between the various organisms that inhabit an ecological system.

For historians, also, the heterogeneity of modern experimental practice poses distinct explanatory challenges. How can the historical development of these experimental systems be understood, their specific temporality represented? One means of doing so is to focus on the progressive stabilization of distinct elements, otherwise known as "black-boxing." When an instrument such as the air pump assumes the status of an accepted means of producing valid phenomena, then it can be said to have become a "black box." The constructivist outlook suggests that this is not simply a matter of the configuration of the hardware, but also involves the creation of a consensus as to how the hardware should be understood. Agreement about the factuality of the phenomenon is, at the same time, agreement that the apparatus has been appropriately used to produce it. (Those who disputed Galileo's astronomical discoveries were also disputing the validity of his telescope; acceptance of the telescope also involved acceptance of the discoveries.) The apparatus thus makes the transition from an object of investigation to an "instrument," properly speaking; henceforth, it is trusted – at least provisionally – to produce authoritative new knowledge. As Andrew Pickering puts it, the experimenters have achieved a stable fit between their model of how the apparatus works and their model of the phenomenon under investigation. Such a stabilization, initially a local and temporary accomplishment, constitutes both the "discovery" of a phenomenon and the black-boxing of a working instrument (Pickering 1989).

One option for historians seeking to recapture the chronological development of modern experimental systems is, therefore, to show how each of their components was black-boxed at some particular time and place. Each part of the assembly can be shown to have traced the path from controversial novelty to reliable instrument, initially supported by a local social consensus and later packaged with other instruments into a hardware ensemble that can be translated to other sites. Such an account could draw profitably upon many features of the sociologists' analyses, for example, Latour's insistence that knowledge and material artifacts are packaged together in the practice of "technoscience" (Latour 1987: 131–132). Arguing along similar lines, Galison talks about the "hardwiring" of theoretical assumptions into the design of experimental instrumentation. He proposes, "We might do well to call these hardwired assumptions 'technological presuppositions' to remind ourselves that machines are not neutral" (Galison 1987: 251).

The historian's interest is not, of course, the same as the sociologist's. The latter may see his or her task as the opening up of black boxes, the recovery of the social processes that contribute to their closure (Latour and Woolgar 1976/1986: 259–260; Pinch 1992). The historian may wish to explore quite different narrative options, involving some kind of accomplishment of closure: to show how a working instrument was made at a particular time, for example, or to examine its function in an unfolding investigative program. In its most narrowly focused form, the history may comprise a story of the creation of a specific phenomeno-technique, that is, the achievement of a stable and reproducible coupling of phenomenon and instrumentation – an embodiment of knowledge in hardware. But phenomeno-techniques have also been shown to become stabilized in the longer term by being embedded within ongoing experimental programs. The black box is sustained as such so long as it can be used profitably in further investigations. As defining features of investigative programs, packaged phenomeno-techniques serve as means of calibrating individual instruments (to determine if they are working properly) and of disciplining experimenters (to assess their competence).

Consider an example from the early nineteenth century: Humphry Davy's discovery of the potential of the voltaic pile as an instrument of chemical analysis. Davy black-boxed the pile by means that included not only laboratory manipulations but also social and rhetorical maneuvers. On the basis of meticulous and persuasively presented experimental work, he dismissed claims that the battery acted to create novel substances, for example when applied to water, insisting that it was essentially an instrument for analyzing a body into its preexisting component parts. By convincingly demonstrating that the device simply decomposed water into hydrogen and oxygen, Davy made it plausible to his contemporaries that the voltaic pile could be used to produce new elements by analysis of other substances. Sodium, potassium, and other elements were soon isolated and the discoveries readily accepted. Establishment of the instrumental efficacy of the pile was, at the same time, the setting up of a standard by which other apparatus could be judged and the crystallization of a community of practitioners who recognized one another's expertise in electrochemistry. Those who had previously claimed to be able to create entirely new substances by the use of electricity were now relegated to the margins of the experimental community. Henceforth, one required a battery of significant power, and a high degree of experimental skill, for one's claims to be taken seriously. Davy's achievement thus played a part in the consolidation of a field of specialist experts, a characteristic development of the second scientific revolution (Golinski 1992a: chap. 7).

Certain difficulties arise when we consider how to assemble stories of this kind together to forge a longer-term account of the development of a complex experimental system. Simply to assume that each instrument was black-boxed independently and assembled successively, in a piece-by-piece fashion, would be to surrender too readily to the temptation to give a whiggish narrative of unproblematic progress. Galison warns us of certain complications (1987: 243–255). He notes that the development of different classes of instrumentation will unfold over different chronological time scales, of which he distinguishes three. The most enduring types of instrument (cloud chambers to detect subatomic particles, for example) remain fundamentally unchanged over relatively long periods of time – several decades at least. Individual variants of these, or more specialized devices, are likely to have shorter lifetimes and more limited experimental applications. On the shortest time scale, every instrument can be said to be under test in the course of each experimental "run." Tinkering and adjustments are always likely to be made on this short-term level. Galison concludes that a history that would do justice to the experimental systems used in modern physics would have to outline developments on at least these three chronological levels and trace interactions between them.

Other approaches to the problem have been suggested by new studies of the development of experimental systems in the biological sciences. Hans-Jörg Rheinberger has considered how experimental systems were crafted to investigate the cellular machinery of protein synthesis in the second half of the twentieth century (1992a, 1992b, 1993, 1994). Rheinberger's vocabulary sometimes verges on the arcane, but his analysis highlights significant features of the temporal unfolding of these systems. He argues that they are characterized by a continual movement of "differential reproduction," so that they constantly yield novel products as they are reproduced. New "epistemic objects" (the objects of experimental knowledge) are created by the use of "technological objects" (those elements of the experimental system that are used instrumentally). Centrifugation at different speeds, for example, was the instrumental means by which a variety of fractions of homogenized cell contents were constituted as "epistemic objects." The fractions were supplied with radioactively labeled amino acids, and the incorporation of these into synthesized proteins was monitored. Rheinberger insists that this process did not merely isolate a fragment of a preexisting reality, separating a "signal" from the background "noise." The instrumental system produced certain entities and destroyed others. Thus, "an experimental system is a labyrinth, whose walls, in the course of being erected, in one and the same movement blind and guide the experimenter" (Rheinberger 1992a: 321).

The constraining force of an experimental system can become apparent to researchers when entities that have been stabilized and used instrumentally revert to the status of objects of investigation. What had been used unproblematically as a tool suddenly demands scrutiny. Thus, the limitations of the ultracentrifuge became clear when it was realized that ribonucleoprotein particles were not segregating neatly into one of the fractions. After the anomaly was investigated, it emerged that the particles contained two distinct subunits, assembled together to form a crucial component of the protein synthesis machinery in the cell cytoplasm (Rheinberger 1992a). This kind of oscillation between instrumental artifact and experimental fact typifies the fluid motion that Rheinberger denotes by "differential reproduction," which enables experimental systems to produce new entities in the course of time.

Robert Kohler has given a more detailed and nuanced view of the development of an experimental system in his recent book on fruit-fly genetics in the first half of the twentieth century, *Lords of the Fly* (1994). In Kohler's story, as in Rheinberger's, the emphasis is laid upon the continuing instability of the elements of an experimental system rather than upon their progressive and irreversible consolidation. This coincides with Kohler's choice of metaphors from the biological sciences instead of cybernetics (the original realm of the "black box") to undergird his narrative. He writes about the symbiosis between the fruit fly, *Drosophila*, and its human hosts, and about an ecology of the laboratories in which both human and fly populations lived. In a quite literal way, Kohler offers an answer to Clarke and Fujimura's call for an ecology of experimental practice.

Thus, in Kohler's view, the fruit fly was *both* a technological artifact *and* a biological organism. On the one hand, it was "engineered" by experimental scientists to serve their purposes, beginning with the pioneering work of Thomas Hunt Morgan and his team at Columbia University in the 1910s. The fly's rapid ten-day reproductive cycle was exploited to breed, through hundreds of generations, precisely the variants needed for genetic experimentation. Specific mutants, for example, which would have died out immediately in wild populations, were maintained because the mutant genes occupied important marker locations on the flies' chromosomes. To map these genes, pure stocks of the mutant strains had to be prepared over many generations, a process Kohler calls "debugging" to eliminate the "genetic noise" that would otherwise obscure the message (Kohler 1994: 66). By tracing the meticulous and laborious procedures by which the fly was made into a tool for genetic mapping, Kohler shows "the construction of *Drosophila* as a standard experimental instrument" (67).

On the other hand, however, the fruit fly remained a living organism with its own adaptive capabilities. Kohler emphasizes how the artificial ecosystem in Morgan's laboratory continued a pattern of long association between the fly and human activity. The *Drosophila* genus, particularly the species *D. melanogaster*, had long tied its fate to human migration, agriculture, and trade. By crossing the threshold into the laboratory, it could be said to be colonizing a new territory and exploiting a new symbiotic relationship with humans. Although its tolerance had certain limits, *D. melanogaster* was well adapted to laboratory conditions; it was hardy, not a fastidious feeder, accustomed to indoor life, and capable of surviving moderate variations of temperature. These were characteristics that had already been selected for in centuries of cohabitation with humans (Kohler 1994: 19–52).

In its dual character as organism and instrument, *Drosophila* shares features with other experimental entities used in the biological sciences. Kohler points out that other living beings, such as the laboratory rat or the bacterium *Escherichia coli*, could be regarded as the core of a similar experimental system. In each case, a "natural" entity has been "engineered" to make more visible and manipulable the features that are of interest to experimenters. But this engineering has not been accomplished without some degree of cooperation with the adaptive capabilities of the organism itself. Each of these organisms could be regarded as having successfully adapted to experimental conditions, as colonizers of the laboratory ecosystem.

In thinking of *Drosophila melanogaster* as an experimental instrument, we have clearly come a long way from Galileo's telescope. The material under consideration and the historical context of its use are obviously very different indeed. There are, however, certain similarities that justify the application of a common theoretical perspective. In each case, we are concerned with the isolation and presentation to view of a certain set of phenomena, which are encountered as distinct from their instrumental framework but also remain closely associated with the instrumentation used to elicit them. In each case, the phenomeno-technique is consolidated within a specific local culture that accords it meaning. And in each case replication of the working instrument in other settings requires reproduction of critical elements of the original context.

When we deal with a living organism we are, of course, more inclined to think in terms of its independent capability for activity, whereas with inanimate objects we tend to invoke a lesser degree of agency, or perhaps only a passive "resistance" to human manipulation. Notwithstanding Latour's insistence on placing human and nonhuman "actants" on the same level, it seems likely that this dichotomy will continue to shape somewhat distinct approaches to the histories of the biological and the physical sciences. In the former case, the capacities of living things for motion, reproduction, and so on, make them particular kinds of allies for recruitment into experimental systems. Nonetheless, the perspective that focuses on the construction of instrumentation, the contribution of specific local cultures to that process, and the continuing rootedness of experimental phenomena in instrumental ensembles can be applied to all experimental sciences. Historians who accept the challenge of this constructivist approach will attempt to identify these general features in particular cases while not ignoring the specific factors of setting and cultural context that differentiate disciplines and historical periods.

As has been mentioned, the temporal dynamics of experimental systems also pose a challenge. Kohler again offers some help here, in his analysis of the transformation of the *Drosophila* system in the 1940s and 1950s. In this period, the fruit fly as an experimental instrument was "reconstructed" in connection with two different experimental programs: studies of the genetics of development, using techniques to transplant tissues between developing larvae; and research on evolutionary genetics, in which populations of wild flies were tracked in the field and laboratory work done to map their chromosomes. The unfolding of each of these programs was governed, according to Kohler, by a dynamic internal to the experimental system. Crucial were "the imperatives of getting the most out of an experimental system, maintaining credibility among fellow experimentalists, and staying ahead of potential competitors. . . . [T]hese practical imperatives were powerful incentives to exploit novel experimental systems opportunistically, whatever one's original intentions" (Kohler 1994: 211).

This interest in following through the internal logic of an experimental system may indeed be a common feature of researchers' adjustment to changing conditions. It can be regarded as a way of protecting the investment of time and resources they have made or, more positively, as an aspect of Pickering's "opportunism in context." But no degree of calculation can determine what will turn up. Kohler recognizes what Rheinberger calls "differential reproduction," the capacity to produce novelty in the course of replication; he writes that "really novel and productive modes of practice not only invigorate but also disrupt the productive and moral economy of working groups" (Kohler 1994: 292). The aptitude of experimental systems to produce novelty continues to challenge experimenters themselves, and their historians.

THE WORK OF REPRESENTATION

Any account of the practices of science has to confront the crucial importance of visual images. In their laboratory ethnography, Latour and Woolgar (1976/1986) observed how instrumental "inscriptions" are routinely transformed into the diagrams and pictures that adorn scientific publications. The tracks of the pens of chart recorders or oscilloscopes are converted into graphs in research papers; and indeed graphs originated historically in connection with the use of self-registering recording apparatus (Tilling 1975). Scientific texts also frequently make use of photography to fix the images of optical instruments, depicting everything from subatomic particle tracks to microscopic organisms to stars and galaxies. Photography invokes commonly shared assumptions that what it shows is real, that the camera cannot lie. The visual impulse in science – the drive for ocular proof, to show how things are, even when those things cannot actually be seen – has also produced many forms of dia-grammatic representation: Michael Faraday's magnetic field lines, sketches of crustal movements beneath the surface of the earth, schematic drawings of how immunological factors fit together, and the many types of maps that are of fundamental importance in the fieldwork sciences (Gooding 1989; Le Grand 1990a; Cambrosio, Jacobi, and Keating 1993; Rudwick 1976). Even the drawing up of data in the form of tables may be said to add an element of visual persuasion to the communication of quantitative findings.

Empirically, visual representation may be studied under several cate-gories: The *settings* in which scientific images are used, from the labo-ratory and the fieldwork site to contemporary textbooks, magazine advertisements, and television, may be scrutinized. In each kind of set-ting, the *function* of images may be reconstructed by investigating the persuasive purposes they are meant to serve and exploring how they are actually interpreted by those who view them. (By analogy to the modes of discursive analysis, described in Chapter 4, we might talk about a "visual rhetoric" and a "visual hermeneutics" in this connection.) Func-tion and setting may also be related to the different *technologies* of visual representation, from drawing and painting to photography and modern electronic media.

Latour (1986) has argued that these factors must all be grasped to-gether. He has suggested that the role of scientific images can be under-stood by focusing on their movement through time and space. It is not the special qualities of particular visual media that are important, in Latour's view, so much as the mobilization of traces that different im-aging technologies make possible. Photographs, maps, printed figures, perspective drawings, all share with bacterial cultures and preserved specimens of plants and animals the character of "immutable mobiles"; all can be moved from place to place while remaining unchanged. They can therefore be used to transport representations of distant phenomena to a single site where they can be manipulated, compared, and combined. For Latour, visual images are a subset of the class of entities that can serve as signs of other, nonpresent entities in the practices of mobiliza-tion and superimposition that constitute science.

Latour's perspective is valuable for its focus on the locations and trans-lations of images, and for its suggestion that their persuasive power is bound up with their capacity for preservation and motion from place to place. But his category of "immutable mobiles" conflates many different kinds of representations and obliterates distinctions between them. A more discriminating analysis might well uncover specific connections be-tween representational techniques and historical settings that Latour has overlooked. A new technology, such as printing, for example, or pho-tography, may make possible a new relationship between the source of images and those who view them. Each innovation in technique may also suggest a different account of the link between the image and the thing represented. Different *philosophies of representation*, that is to say, may be linked to the use of different visual media. But, equally, different representational conventions may be adopted for the use of the *same* medium. Drawings, for example, may answer to the demand for an idealized representation of a type of phenomenon, or for naturalistic or realistic depiction of an individual specimen. Photographs, viewed individually, tend to incorporate a certain amount of individuating naturalistic detail, but those chosen to appear in a publication may be selected to represent typical features of a class of objects. Different notions of objectivity may be implicated in such a choice of stylistic convention (Daston and Galison 1992).

Constructivism suggests that these issues can be approached historically by tracing the events surrounding the introduction of new techniques. In contemporary research practice, representational techniques are superimposed upon one another. For example, an electron micrograph may be assembled in a montage, labeled, and then selectively redrawn to isolate particular features (Lynch 1985). It has been suggested, in fact, that the superimposition of heterogeneous methods of image-making is a telling feature of experimental practice in modern science (Lynch and Woolgar 1990: 2). As with the arrays of instruments encountered in the contemporary laboratory, history offers a means of disentangling the complexity. Each method of representation can be documented at its time of origin, and the arguments surrounding its initial use recovered. In this way, we recapture the work of construction and interpretation that goes into forging a new representational technology, the forgotten labor that makes it possible for a new type of image to come to seem natural.

To adopt this perspective is not to challenge the adequacy of particular images as depictions of reality, but simply to set the judgmental question aside in the interest of grasping the functions and meanings of different kinds of images. Even images that we now regard as woefully inaccurate, such as medieval anatomical drawings or maps, deserve to be studied sympathetically in relation to their meaning for those who created and used them. Conversely, photography, a medium we now accept as a window onto the real world, requires historical scrutiny to uncover the circumstances in which it came to be accorded that status.

As with language and instrumentation, we are inclined to trace the origins of scientific visual images to the seventeenth century, when the Renaissance inventions of the printing press and single-point perspective were firmly established as representational technologies. Boyle included an engraved drawing of the air pump in his *New Experiments Physico-Mechanical* (1660). As Shapin points out, the conventions employed in the

Figure 6. Robert Boyle's air pump, as it was shown in the frontispiece to his *New Experiments Physico-Mechanical* (1660). Reproduced by permission of the Syndics of Cambridge University Library.

depiction paralleled those implicit in Boyle's rhetoric. The pump was individuated and presented in specific circumstances. The hatched shading was designed to indicate a single three-dimensional object, in a particular setting, unlike the schematic outlined apparatus typically shown in twentieth-century scientific illustrations (Shapin 1984: 491–492). Beyond indicating these parallels with Boyle's "literary technology," Shapin does not pursue the question of the form of representational practice involved here – the assumptions made about how illustrations communicated with their viewers, or the way in which artistic conventions reflected those assumptions. Boyle, indeed, might not be the best subject to consider in the attempt to answer these questions.

Mary Winkler and Albert Van Helden have taken the matter further in discussions of visual images in early-seventeenth-century astronomy (1992, 1993). They note that the invention of the telescope in the early years of the century created problems of credibility for those who sought to communicate their observations with the novel instrument. In his *Siderius Nuncius* (1610), Galileo went some way toward addressing this problem by developing techniques to represent the moon visually. He made ink-wash drawings, from which copper engravings were subsequently prepared and printed. But the burden of proof was placed more heavily on Galileo's words than on his pictures: The crucial phenomena were given lengthy verbal descriptions, and the plates departed from naturalism by exaggerating the features Galileo wanted to emphasize. Winkler and Van Helden suggest that techniques of naturalistic representation were not consistently used in astronomy until the 1640s, when Johannes Hevelius published his monumental *Selenographia* (1647). Hevelius, who undertook the engraving of the numerous plates himself, included images of all the planets, along with the sun and stars and a series of unprecedentedly detailed maps of the moon. He also enclosed a portrait of himself and pictures of his telescopes and his lens-grinding equipment. As in the case of Boyle a few years later, one can plausibly conclude that Hevelius was seeking to induce some kind of "virtual witnessing" among his readers, permitting them both to see what was seen through the instrument and to imagine themselves present at the site of the observations.

A paradigmatic instance of the use of visual images for scientific communication in the seventeenth century is Robert Hooke's *Micrographia* (1665). Subtle readings of Hooke's text and its accompanying plates have been given by Michael Aaron Dennis (1989) and John T. Harwood (1989). Dennis argues that some of Hooke's discursive moves echoed Boyle's literary technology, but that his reliance on visual representation gave his work a quite different overall effect. Hooke's textual commentary described in detail the ways he prepared his specimens and the lengthy manipulations to which he subjected them before making his drawings.

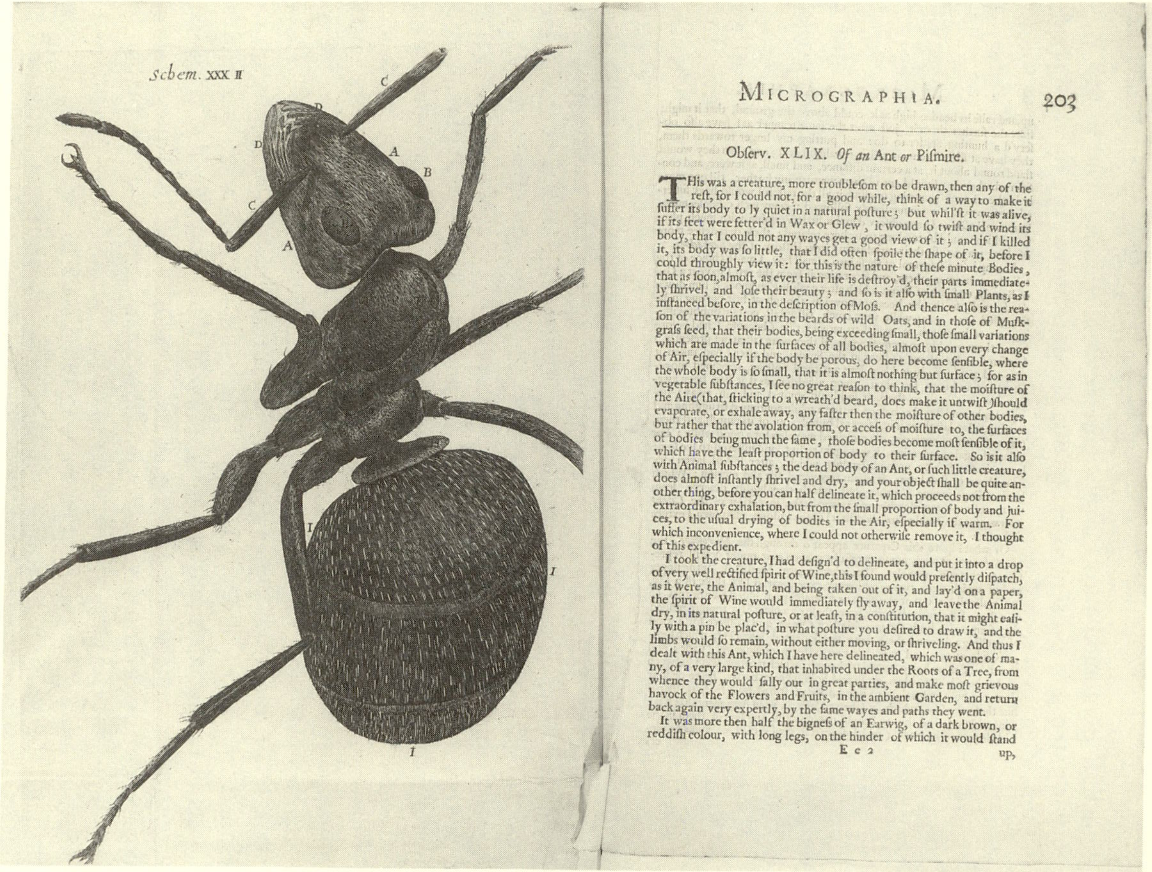

MICROGRAPHIA. 203

Obſerv. XLIX. *Of an* Ant *or* Piſmire.

THis was a creature, more troubleſom to be drawn, then any of the reſt, for I could not, for a good while, think of a way to make it ſuffer its body to ly quiet in a natural poſture; but whil'ſt it was alive, if its feet were fetter'd in Wax or Glew, it would ſo twiſt and wind its body, that I could not any wayes get a good view of it; and if I killed it, its body was ſo little, that I did often ſpoile the ſhape of it, before I could throughly view it: for this is the nature of theſe minute Bodies, that as ſoon, almoſt, as ever their life is deſtroy'd, their parts immediately ſhrivel, and loſe their beauty; and ſo is it alſo with ſmall Plants, as I inſtanced before, in the deſcription of Moſs. And thence alſo is the reaſon of the variations in the beards of wild Oats, and in thoſe of Muſk-graſs ſeed, that their bodies, being exceeding ſmall, thoſe ſmall variations which are made in the ſurfaces of all bodies, almoſt upon every change of Air, eſpecially if the body be porous, do here become ſenſible, where the whole body is ſo ſmall, that it is almoſt nothing but ſurface; for as in vegetable ſubſtances, I ſee no great reaſon to think, that the moiſture of the Aire (that, ſticking to a wreath'd beard, does make it untwiſt) ſhould evaporate, or exhale away, any faſter then the moiſture of other bodies, but rather that the avolation from, or acceſs of moiſture to, the ſurfaces of bodies being much the ſame, thoſe bodies become moſt ſenſible of it, which have the leaſt proportion of body to their ſurface. So is it alſo with Animal ſubſtances; the dead body of an Ant, or ſuch little creature, does almoſt inſtantly ſhrivel and dry, and your object ſhall be quite another thing, before you can half delineate it, which proceeds not from the extraordinary exhalation, but from the ſmall proportion of body and juices, to the uſual drying of bodies in the Air, eſpecially if warm. For which inconvenience, where I could not otherwiſe remove it, I thought of this expedient.

I took the creature, I had deſign'd to delineate, and put it into a drop of very well rectified ſpirit of Wine, this I found would preſently diſpatch, as it were, the Animal, and being taken out of it, and lay'd on a paper, the ſpirit of Wine would immediately fly away, and leave the Animal dry, in its natural poſture, or at leaſt, in a conſtitution, that it might eaſily with a pin be plac'd, in what poſture you deſired to draw it, and the limbs would ſo remain, without either moving, or ſhriveling. And thus I dealt with this Ant, which I have here delineated, which was one of many, of a very large kind, that inhabited under the Roots of a Tree, from whence they would ſally out in great parties, and make moſt grievous havock of the Flowers and Fruits, in the ambient Garden, and return back again very expertly, by the ſame wayes and paths they went.

It was more then half the bigneſs of an Earwig, of a dark brown, or reddiſh colour, with long legs, on the hinder of which it would ſtand

Ee 2 up,

Figure 7. An opening from Robert Hooke's *Micrographia* (1665) showing the complementary relationship between text and image. Hooke's treatment of the ant is richly detailed in the prose narrative and in the quality of the visual image. Reproduced by permission of the Syndics of Cambridge University Library.

These precautions – dousing an ant with alcohol, for example, to make it stay still, or trying the effect of different lighting conditions on the appearance of a fly's eye – were described to convey an impression of Hooke's careful experimental protocols, and hence to convince readers that he could serve as a true medium to convey natural phenomena. In large measure, it was these textual details that Hooke deployed to persuade his readers of his "sincere hand and faithful eye" (Dennis 1989: 323, 343–344; Harwood 1989: 138, 143–144).

The 246 pages of text were, however, only part of the work; they were complemented by 38 pages of plates. Specific resources were offered by the representational techniques available to Hooke, and specific problems were posed. The large plates, in which naturalistic shading and shadows were used to convey a three-dimensional appearance, and in which minutiae of texture and surface features were meticulously rendered, presented familiar objects such as insects in a dramatically unfamiliar way. At the same time, however, homely analogies were resorted to – seeds of thyme viewed under the microscope, for example, were likened to a dish of lemons in a painting. This oscillation between the realms of the familiar and the exotic was frequently repeated: The mold growing on a piece of stale bread was shown in an unfamiliar view but described by invoking the analogy of flowers blowing in the wind. The application of naturalistic conventions and mundane analogies to represent an alien yet intimately close world was striking, and perhaps also unsettling (Harwood 1989: 138–142; Alpers 1983: 73–74, 84–85).

Hooke also had to address the issues of engraving and printing, the technologies that mediated between his own drawings and their public appearance. On the one hand, he reassured his readers that these techniques could be regarded as entirely transparent – he was said to have closely directed the engraving, which was performed by unnamed artists who did as they were told. On the other hand, attention could be drawn to the work of the engraver when Hooke wanted to identify a mistake in translating from his drawings, rather as Boyle mentioned his laboratory technicians when he wanted to blame them for errors (Dennis 1989: 314–315; cf. Shapin, 1994: chap. 8). Hooke retained the freedom to open the gap between the original drawing and the subsequent engraving by specifying that he had written his commentary after seeing the finished plates. He thus located himself in the place of the reader, whom he guided in viewing through the illustrations to the reality that lay beneath them (Dennis 1989: 344).

Underlying these practices was a rationale that Dennis labels a "hermeneutic of representation." The microscope, even while it testified to the power of human art, strengthened the distinction between products of art and products of nature. This was driven home by Hooke in his first observation of the point of a needle. When applied to such an artificial object, the microscope revealed only the imperfections of human art and the limits of human design; but when applied to natural objects like a fly it disclosed the unsuspected intricacies of divine workmanship. God's design was shown to be infinitely superior to man's, and it was to be revealed by the instruments He had granted mankind. God had inscribed in the natural world His message of design, the "micrographia" or "small writing," to be read by the microscope. Hooke wrote: "who knows but the creator may, in those characters, have written and engraven many of the most mysterious designs and counsels, and given man a capacity, which assisted with diligence and industry, may be able to read and understand them" (quoted in Dennis 1989: 336).

This hermeneutical strategy was contrasted by Hooke with the attempts of cabalists to decode a divine message supposedly concealed in the Hebrew language. The pretensions of these misguided "Rabbins" were exposed by the microscope itself. Hooke recorded that, when turned on printed or written script, the instrument revealed the crude imperfections typical of human fabrications. While the cabalists mistakenly expected truth to be revealed through the enigma of a linguistic code, experimentalists understood that God had written His message in the natures of things themselves. This was the original language of Adam, to be recovered by invention and instrumentation. As Dennis puts it:

> Although the Fall rendered man's senses, memory, and reason imperfect, man could repair much of the damage to the senses by using the new adjuvant technologies appropriately and seeing more evidence of God's power, as well as what God may have written into creation itself. Disciplined seeing worked to produce transparent observers whose microscopical and telescopical observations captured "the things themselves"; representation worked as an interpretive method through the claim that there was no processing of the object, only the *re-presentation* of the thing itself. (Dennis 1989: 337; cf. Alpers 1983: 93)

With this account of Hooke's "hermeneutic of representation," Dennis gestures toward a much broader interpretation of the reshaping of representational techniques in the early-modern period. Michel Foucault (1966/1970) discussed the shift from a Renaissance "episteme" to the succeeding "Classical" episteme in the mid-seventeenth century. The former regime of knowledge was dominated by the relationship of *similitude*, in which analogies and correspondences between entities in different cosmic realms were multiplied in endless profusion. The latter was dominated by the figure of *representation*, which acknowledged an ontological gap between words and things but sought to bridge it by the relationship of signification. Under the aegis of representation, visual images are distinguished from discourse, in that they exploit a resemblance between objects and their depictions, rather than the arbitrary semiotic relationship that holds between words and things. On the other hand, images share with words their status as copies of an original that is supposedly prior to and independent of them (Foucault 1983: 32–33).

Michael Lynch and Steve Woolgar have suggested that this offers a clue to the understanding of representation in contemporary scientific practice. Exploiting the endless possibilities of resemblance, scientists superimpose image upon image, progressively isolating and clarifying the features they seek to emphasize. Having done this, they are able to present the final image as simply a transparent copy of the original, or (as Dennis puts it) a "re-presentation of the thing itself." Chains of images are "laboriously built up and then 'forgotten' in the presumptive adequacy of their reference to an 'original' " (Lynch and Woolgar 1990: 7).

If this is part of the story, it cannot be the whole story of the various techniques and functions of scientific images since the seventeenth century. Leaving aside the well-canvassed difficulties of Foucault's periodization, and his invocation of excessively abrupt dislocations in historical time, his account of visual representation may also be deepened and variegated. Adequacy of reference to an original may well be sought in all scientific images, but notions of what constitutes the original and how it can be adequately portrayed may vary, as a comparison with different styles in art would suggest. It may also be true that those who create images in the experimental sciences generally seek to efface their own role, to render themselves – along with the media they employ – transparent instruments of representation. But the means by which this is accomplished may be expected to differ with different visual technologies. The photographer is not implicated in the production of his or her images, for example, in quite the same way that the artist is, though the photographer's role is undeniably important.

These reflections are borne out by the intriguing analysis by Lorraine Daston and Peter Galison (1992) of images from a range of printed texts drawn from the natural and physical sciences since the seventeenth century. The authors discuss these images in relation to the notions of objectivity that they embody. They argue that a significantly new conception of objectivity was introduced in the nineteenth century which linked objective representation with a capacity for discipline and self-restraint on behalf of the observer. The observer was expected to practice a heroic asceticism, laboriously attained, in order to keep at bay the temptations of subjective judgment. He – and the model observer was distinctly masculine – was required to adopt the morality of the machine (Daston and Galison 1992: 81–84).

Prior to the emergence of this ideal, questions about the accuracy of scientific images were focused on the issue of their "truth to nature." In the seventeenth and eighteenth centuries, it was thought that this could be achieved only when the idiosyncrasies of individual specimens were eliminated. Those who wanted to portray, for example, human anatomy or the forms of plants could therefore choose between two options. Either an "ideal" form could be depicted, representing a degree of perfection not found in any actual specimen, or a "characteristic" example could be shown, in which the features typical of a class as a whole were located in a selected individual. The atlas of human anatomy by Bernhard Albinus, published in 1747, pursued both strategies but subordinated the second to the first. Albinus's meticulous drawings were prepared from a carefully chosen skeleton, "very well proportioned; of the most perfect kind, without any blemish or deformity." But this characteristic specimen was further improved upon in the drawings to remove features that fell short of a notional ideal, "so those things which were less perfect, were mended in the figure, and were done in such a manner as to exhibit more perfect patterns" (quoted in Daston and Galison 1992: 90). As Daston and Galison point out, Albinus saw no contradiction between the aims of ideal perfection and exact representation; for him, "truth to nature" required depiction of the universal in preference to the particular.

Declining a Foucaultian rigidity of periodization, Daston and Galison also remark that the eighteenth century showed some instances of what later became the prevailing mode of visual representation, in which the individuating features of particular specimens were naturalistically depicted. They point to the case of the surgeon William Hunter's *Anatomy of the Human Gravid Uterus* (1774). Hunter's thirty-four engraved plates were distinguished by a high degree of naturalistic realism, with sharp edges to the features depicted, subtle shading to indicate natural lighting, and surface textures rendered with extraordinary detail. Viewed by the modern eye, accustomed to the conventions of photography, Hunter's plates appear dramatically and disturbingly photographic, their unsettling qualities being amplified by the subject matter – the dissected bodies of pregnant women and their fetuses. Indeed, as Ludmilla Jordanova (1985) has argued, Hunter's pictures, though sharing certain conventions with other contemporary representations of the womb and the fetus, are distinguished by their blend of naturalistic detail and pitiless mutilation of the female body.

For Hunter, the choice of naturalistic representation of a particular specimen was a deliberate one, reflecting a preference for the object "represented exactly as it was seen" over one shown as "conceived in the imagination" (quoted in Daston and Galison 1992: 91–93). He admitted, however, that naturalism was an achieved effect, requiring some degree of intervention to prepare the corpse by injecting the womb with spirits and the blood vessels with colored wax. Only with this kind of manipulation could all the relevant features be made visible and what Hunter called "the mark of truth" conveyed (Jordanova 1985: 394).

Hunter's atlas is hailed by Daston and Galison as evidence that "sci-

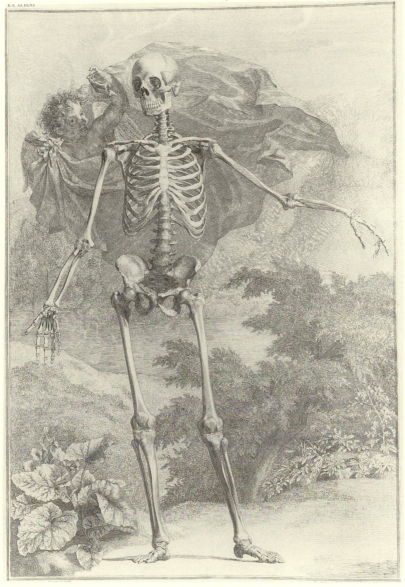

Figure 8. A representative plate from Bernhard Albinus's anatomical atlas. Daston and Galison (1992) note that, in Albinus's plates, a typical human skeleton was modified to produce an image more in accordance with a notional ideal. From Albinus, *Tabulae sceleti et musculorum corporis humani* (1747), plate 1. Reproduced by permission of the Syndics of Cambridge University Library.

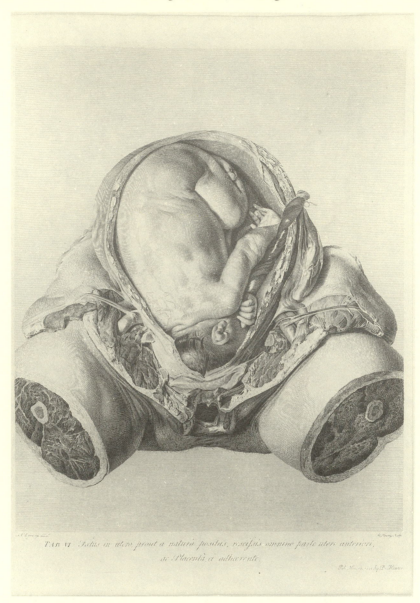

Figure 9. A representative plate from William Hunter's anatomical atlas. Daston and Galison (1992) and Jordanova (1985) have commented on Hunter's use of abundant naturalistic detail in his starkly realist images of the dissection of a pregnant female corpse. From Hunter, *The Anatomy of the Human Gravid Uterus* (1774), plate 6. Reproduced by permission of the Syndics of Cambridge University Library.

entific naturalism and the cult of individuating detail long antedated the technology of the photograph" (1992: 93). It takes its place in a series of stylistic innovations that prepared the way for photography. The technical invention, when it arrived in the 1830s and 1840s, was assimilated into a particular order of representational practice, in which naturalistic detail was already established as the "mark of truth." Hence, surprising elements of continuity bridge the chasm of photography's invention. In the 1730s, William Cheselden was already using a camera obscura to prepare drawings of a skeleton. In the late nineteenth century, anatomists were still making drawings from photographs because the latter, on their own, could not show important features with sufficient clarity. For Daston and Galison, the critical issue is not the technology of the camera as such but the way in which representational practices are tied to a moral order of objectivity: "What we find is that the image, as standard bearer of objectivity, is inextricably tied to a relentless search to replace individual volition and discretion in depiction by the invariable routines of mechanical reproduction" (1992: 98).

The critical feature of the photograph, on this reading, was not its realism but its production by a mechanical process that could be regarded as largely independent of human intervention. This was indeed a commonly observed aspect of early photography, and probably of key importance in making it acceptable as a means of image-making. William Henry Fox Talbot, the inventor of the negative/positive process and the calotype, continually emphasized the character of photography as a practice that allowed nature to create pictures automatically. Speaking to the Royal Society in 1839, Talbot described how he had made a series of images of his house in the country, which was thus "the first [building] that was ever yet known *to have drawn its own picture*" (quoted in Snyder 1989: 12). The notion of a process by which nature inscribed its own representation lay behind the coining of the word "photograph" by John Herschel, who presented the results of his own experiments in 1839, and was further emphasized by Talbot's significantly entitled book, *The Pencil of Nature* (1844–1846). Edgar Allan Poe later subjected a photograph to microscopic examination and recorded that, unlike works of human art, closer scrutiny disclosed unsuspected detail and "a more absolute truth, [a] more perfect identity of aspect with the thing represented" (quoted in Daston and Galison 1992: 111). According to the criteria laid out in Hooke's *Micrographia*, one would have to conclude that the photograph was not a human artifact at all. Allowing nature to represent itself, it seemed to eliminate human agency, producing an image whose truth resided in the preservation of redundant but highly particularistic detail.

This being so, Daston and Galison are right to point to the place of photography among a cluster of techniques of automatic graphical representation developed in the nineteenth century. By the 1870s, the physiologist Etienne Jules Marey was hailing such "inscription instruments" as the kymograph and the camera as yielding "the language of the phenomena themselves" (quoted in Daston and Galison 1992: 116). As Simon Schaffer has noted, self-registering instruments of this kind had come to replace the bodily senses of the investigator or his audience, which had been problematic but pervasive resources for eighteenth-century experimentation (Schaffer 1994b). Within the moral economy of nineteenth-century scientific disciplines, photography was just one of the technologies of automatic representation harnessed to the pursuit of a mechanical ideal of objectivity.

A fine case study of the way photography worked in this context has been done by Holly Rothermel (1993). She considers photographic studies of the sun, from the 1850s to the 1870s, showing how Warren De la Rue and George Biddell Airy tried to overcome the subjectivity of individual human observers with the mechanical objectivity of the photograph. De la Rue, an English businessman and prominent astronomer, used the wet collodion plate process for astronomical applications and recorded his conviction that scientific photography "in its very principle carries with it all extinction of individual bias" (quoted in Rothermel 1993: 137). That photography should have appealed also to Airy, a Cambridge professor and the Astronomer Royal from 1835 to 1881, is not surprising in view of his enthusiasm for technical innovations to eliminate the biases of individual observers. Airy, whose factory-like discipline and use of self-registering devices to calibrate observers' reaction times have been mentioned in Chapter 3, praised photography as a contribution to making astronomy "self-acting" (Schaffer 1988; Rothermel 1993: 144, 153).

These hopes were vested initially in photographic studies of solar eclipses. The eclipses of 1836, 1842, and 1851 had generated a series of controverted observations concerning such phenomena as "Baily's beads," red flares, and the solar corona, which all attempts at standardization were unable to reconcile. In 1860, De la Rue succeeded in taking a series of photographs of a solar eclipse that exhibited in a generally acceptable form some of the previously contested phenomena. Airy hailed the achievement and, in a presumably unconscious echo of Poe's procedure, suggested that the prints be subjected to microscopic measurement to extract quantitative information. Less successful was the subsequent attempt to use photography to improve measurements of the timing of the transit of Venus across the face of the sun in 1874. In this case, the photographic plates proved unable to supply a conclusive result. Enlargement rendered the edges of the sun and the planet indistinct, and different analysts reported different measurements. Significantly,

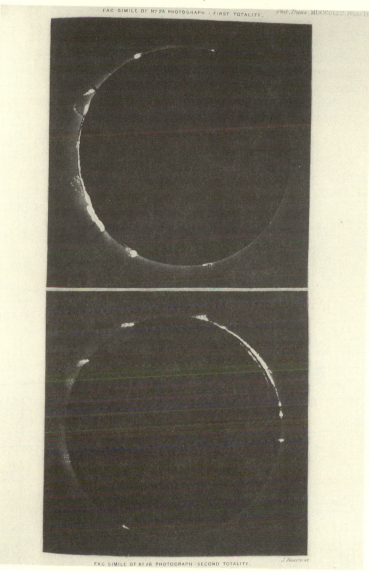

FAC SIMILE OF Nº 25 PHOTOGRAPH - FIRST TOTALITY.

FAC SIMILE OF Nº 26 PHOTOGRAPH - SECOND TOTALITY.

Figure 10. Two of Warren De la Rue's photographs of a total solar eclipse in 1860. Rothermell (1993) has discussed the early use of photography to study eclipses, and the problems of interpretation of images such as this. Mezzotint facsimile from the *Philosophical Transactions* (1862), plate 9. Reproduced by permission of the Syndics of Cambridge University Library.

Airy blamed, not the apparatus but the personal biases of the individuals responsible for the measurements. He wrote to De la Rue:

> Our difficulties do not arise from anything which suggests suspicion that the apparatus failed. . . . [T]he great difficulty is the discordance between the measures of the same space made by different observers at different times. The discordance between two able men, Captain Tupman and Mr Burton amazes me. (quoted in Rothermel 1993: 167)

Airy's judgment exhibits a symptomatic faith in the ability of photography to serve scientific representation. As a process by which natural phenomena registered their own images, its fundamental validity seemed to be beyond question. If there was a problem, it had to be in humans' failure to standardize their interpretations of photographs, not in any limitations of photographic representation itself. For Airy, such a judgment was of a piece with his ambitions to make scientific investigation partake of mechanical objectivity and to reduce investigators to the role of machines. But the judgment was not a purely idiosyncratic one; it reflected a widespread assumption, which remains prevalent today, that photography is (or can become) a transparent medium of representation, that it can be "forgotten in the presumptive adequacy of its reference to an original."

Bearing this in mind, we might be tempted to identify photography as the paradigm technique of scientific representation, its images the ultimate "immutable mobiles." We could point to the prevalence of photographs in many fields of scientific research, to the pervasiveness of the descendant technologies of film and television, and to their vital role in extending the audiences for science beyond the circle of specialist practitioners. But the constructivist outlook would suggest that we hesitate before embracing such a conclusion, and remember instead the manufactured status of photographic images. Three points seem worth emphasizing. First, historical studies of the origins of photography have strongly argued that it was far from a natural process, that its images depended on a certain context – cultural, technological, and social – for their acceptance, and that they remained problematic in many applications. Even enthusiastic devotees, such as De la Rue and Airy, had to labor to make photography work for them in the way they wanted, and with mixed success.

Second, although the role of photography in certain realms of scientific practice is now moderately secure, we should be alert to the possibility that newly interesting phenomena or new representational tasks might call that role into question. Homer Le Grand has described how photographs have been displaced in recent decades from their previous prominence in textbooks of physical geology (1990a: 242–243). Whereas authors used to draw upon photographic illustrations very frequently to emphasize surface landforms, the new concentration since the 1960s on continental drift has required different kinds of pictures. Reconstructive maps of moving continents and shifting magnetic poles, schematic diagrams of seafloor spreading and plate tectonics, which played a vital role in the debates over the drift hypothesis, have now been lifted into the textbooks. Le Grand argues that they were "not merely illustrative but constitutive of the versions of Drift which eventually became the new orthodoxy" (243). Photographs simply could not have satisfied the need for specifically theory-laden images to show geological processes that are actually invisible to the human eye.

Finally, with respect to the role of photography and related technologies in the popularization of science, a constructivist perspective would suggest that we ask exactly what messages are being communicated by these means. Rather than presuming that photography serves as an entirely transparent medium of representation, we should ask how its meanings are created by those who produce the images and by those who view them. Photographs mounted in natural history museums, or television documentaries on animals in their natural habitats, present themselves as windows onto the world of nature; but a constructivist analysis cannot remain satisfied with that construal. Quite novel, even unanticipated, meanings might in fact be found to be communicated by museum displays or natural history films. The technologies of representation involved have to be rendered opaque and scrutinized for the part they play in the human labor of making meaning.

6

Culture and Construction

We say that the laws of Newton may be found in Gabon and that this is quite remarkable since that is a long way from England. But I have seen Lepetit camemberts in the supermarkets of California. This is also quite remarkable, since Lisieux is a long way from Los Angeles. Either there are two miracles that have to be admired together in the same way, or there are none.

Bruno Latour, *The Pasteurization of France* (1988a: 227)

THE MEANINGS OF CULTURE

This book has traced the descent of recent work in the history of science from the radical theoretical outlook of the Strong Programme and the sociology of scientific knowledge. I have proposed that the sociologists' arguments against traditional epistemology opened up a wide range of issues to renewed historical investigation. The ways in which experiments were understood, for example, or theories constructed, became issues for empirical research. The widespread success of the sciences in their technological manifestations seemed to call for fresh analysis. Once the assumption that science is inherently universal was challenged, then historians could start to examine the material and cultural means by which the global extension of scientific knowledge has been constructed, to the degree that it actually has.

Not everyone will agree with the filiation I have been tracing here. Other quite plausible genealogies might be outlined that would link recent historical studies of science to quite different philosophical or disciplinary approaches. Some sociological studies of science have been inspired by alternative theoretical perspectives, such as ethnomethodology or symbolic interactionism. Anthropology has exerted an attraction for researchers in many areas of historical inquiry, including the history of science. Feminism has had an enormous influence in shaping many academic disciplines, not least in raising the categories of gender and sexuality to prominence in the work of many historians. The existence of a "new cultural history" has recently been proclaimed, drawing upon these roots and other trends in academic thought, such as semiotics and deconstruction (Hunt 1989). In light of this, some historians of science have sought to orient the field as part of a wider "postdisciplinary" movement, adopting the label "cultural studies of science" or "cultural history of science."

This is an unobjectionable, if imprecise, label for a wide variety of studies that have come onto the scene in the last few years. Many of these have developed lines of inquiry that were not explored by SSK, concerning, for example, the heterogeneous practices that comprise scientific research, its discursive traffic with other cultural realms, or the formation of the identities of scientific practitioners. In my account, I have tried to indicate how the path opened up by the sociologists can lead into these territories. I have indicated, for example, how anthropological and feminist perspectives might help us to understand the construction of identities among early-modern scientific practitioners. Repeatedly, this book has traced a journey from the foundations of constructivism to the historical research that is building upon them, frequently in ways the pioneers could not have anticipated. We have, in effect, followed the path of history of science from its rebirth in the philosophical disputes of the 1960s to its current situation amid the broad stream of cultural studies of science (cf. Pickering 1995a: 217–229).

And yet the term "cultural," which gives us a convenient label, has certain problematic resonances. "Culture" is a tricky thing to define, but the term seems to be readily understood, perhaps because it sometimes invokes particular ideological connotations. These seem to reverberate in Joseph Rouse's use of the label "cultural" to demarcate certain recent studies from those aligned with SSK or "social constructivism" (Rouse 1993b). Used in this way, the word seems to be serving to exclude undesirable approaches, which are labeled "social" or "sociological" and are contrasted unfavorably with the "cultural" (Dear 1995a). In part, the opposition works by playing upon the bad odor in which the word "social" has come to be held among conservative thinkers in the last decade or two, for whom it is uncomfortably closely associated with "socialism." But the ideological functions of "culture" as an analytical term are more deeply rooted than that. It is worth considering the issue for its bearing upon our understanding of science.

The literary critic Raymond Williams, regarded as one of the founders of contemporary cultural studies, devoted himself to revealing the historical connotations of the term "culture," which he traced back to newly industrialized England in the early nineteenth century. He stressed that the term long retained a moral and aesthetic dimension which it had inherited from this period. Conservative writers in particular used "culture" to designate both a description of intellectual activities, separate from the social relations characteristic of emergent industrial society, and a standard for moral evaluation of those activities. As Williams put it, "culture" in the works of such writers as the poet and educationalist Matthew Arnold named an abstraction and an absolute. Even when purportedly used in a descriptive sense, the word carried with it suggestions of the possibility of judgment according to standards that were set above the world of contingent social formations. The dual descriptive and evaluative meaning of the term was retained as it expanded its scope to cover a whole way of life, a framework for understanding common experience. In the twentieth century, T. S. Eliot resuscitated the evaluative connotations of the term and the sense of intellectual activity sundered from social relations. Thus, Eliot revived the rhetorical maneuver by which "culture became the final critic of institutions, . . . yet it was also, at root, beyond institutions" (Williams 1963: 136).

The notion of culture as a framework of values beyond the realm of the social was brought to bear upon science in C. P. Snow's influential lecture on *The Two Cultures and the Scientific Revolution* (1959/1993) and in the notoriously intemperate response to it by the Cambridge literary critic F. R. Leavis (1963). In his lecture, Snow voiced what was obviously a deeply felt conviction (which he claimed was shared by many other scientists) that science and technology offered the best hope for satisfying human needs. In the face of the tragedy of individual death, he explained, science offered "social hope," the consolatory knowledge that one was working to alleviate the unnecessary sufferings of humanity by increasing its material well-being (Snow, 1959/1993: 6–7, 84–86). This conviction, a kind of secular displacement of religious aspirations, linked Snow back to the radical scientists with whom he had associated at Cambridge University in the 1930s. It remained, however, curiously unconnected with any substantial analysis of scientific institutions or more than anecdotal evidence of scientists' attitudes. Snow's characterization of science as a culture was certainly imbued with moral and aesthetic values, but these were quite detached from consideration of its context in concrete social relations. Tellingly, his recipe for diffusing the beneficial results of science to relatively less favored nations was simply to plant "alpha-plus" scientists and engineers like seeds in the soil of the developing world.

Leavis offered nothing better. He viciously challenged Snow's credentials as a scientist and a novelist, claiming that he was an incompetent informant on the profundities of scientific knowledge and the human condition. As far as Leavis was concerned, Snow had failed to convince his readers that there *was* a scientific culture worthy of comparison with that to be found in great literature – or at least he had failed to convince them that he could speak with authority about it. In a way, this was nothing but the reverse aspect of Snow's fundamental point: that the scientific culture was divided by a deep gulf from the entrenched literary one, which held sway over the leading minds of the Western world.

Leavis could be taken as validating Snow's claim that "literary intellectuals" despised science and its practitioners because they could not recognize the moral values that inspired their vocation.

Although Leavis and Snow took opposite views as to whether science measured up as a culture, they operated with a common notion of what a culture was. For both writers, "culture" denoted a scheme of moral and aesthetic values that was quite independent of the social conditions framing intellectual life. In the early 1960s, it seems, analysis of science as a culture could make no further progress than this. The term "culture" was so heavily tinged with its evaluative connotations that Snow did not consider the need to tie it more firmly to social reality. And, for Leavis, the claim that science could convey its own scheme of values was so preposterous as not to deserve serious consideration. The understanding of science as a culture remained hampered by the aesthetic and moral connotations of the crucial analytical term, and by the assumption that a framework of values was something apart from the realm of the social.

Current uses of the terms "culture" and "cultural" in connection with science cannot remain unmarked by this history. It resurfaces whenever these words are used to draw attention away from issues of social relations. But, of course, the cultural does not have to be set up as a polar opposite to the social. And, although Rouse might seem to be doing this, his actual definition of "culture" is inclusive, rather than exclusive, of "social practices" (Rouse 1993b: 2). The development of cultural studies after the 1960s has offered plentiful resources for those who want to integrate an analysis of science as a cultural formation with consideration of its embeddedness in the social world. Much of this work has taken its point of departure from Williams's pioneering historical archaeology of the term "culture" and his reappropriation of it for the intellectual left. Commenting in 1986 on the legacy of the Snow/Leavis dispute, Williams wrote that the controversy had been:

> generous and passionate in its intended motives, but hopelessly confused by its enclosing categories and by its consequent omission of that whole area of practice and learning, itself reduced in shorthand to "society," where the real and complex relationships are continually generated, tested, amended and renewed.
>
> It is a substantial achievement of the generation which inherited that kind of misdirection that ways have been found of addressing the undoubted problems more precisely and more specifically. (Williams 1986: 11)

Among the new methods Williams alluded to, he specifically focused on those that analyzed scientific discourse as a semantic system, the elements of which shared meanings with other discursive fields. A "close verbal analysis" of these formations, he wrote, would be "inevitably social and historical"; it could illuminate the transfer of concepts, metaphors, and models between scientific and other discursive realms. This kind of study would be resisted, Williams conceded, by traditional literary critics, as well as by "the simplest exponents of 'science' as a wholly autonomous form of knowledge" (1986: 11–12).

Literary and linguistic methods of this kind (which I have surveyed in Chapter 4) have been complemented by other approaches, derived particularly from anthropology, which have also attempted to analyze cultural formations as systems of meaning in specific social settings. Meaning, in this work, is viewed as the outcome of social action as much as of discourse itself. The most articulate defender of this approach has been the anthropologist Clifford Geertz. Culture, Geertz asserts, is an "acted document," of which one asks, not "what causes such behavior?" but "what is being said?" Action, whether ritualized or apparently spontaneous, can be interpreted as "social discourse" (Geertz 1973: 10, 18). The task of the ethnographer, or any student of human behavior, is thus analogous to that of the textual scholar: to acquire the skills of interpretation that allow one to discern the meanings actions have within the context in which they are produced. In a much cited formulation of his views, the essay on "Thick Description: Toward an Interpretive Theory of Culture," published in 1973, Geertz specified how his outlook arose from "an enormous increase in interest, not only in anthropology, but in social studies generally, in the role of symbolic forms in human life. Meaning, that elusive and ill-defined pseudoentity we were once more than content to leave philosophers and literary critics to fumble with, has now come back into the heart of our discipline" (Geertz 1973: 29).

Viewed in this way, as the hermeneutic enterprise of "reading" human cultures, the social sciences (which, Geertz recalls, comprise "the 'Third Culture' Snow forgot" [1983: 158]) have opened up new paths toward an understanding of science in general. Scientific communities can be approached, like any others, by ethnographers aiming to discern the meanings of their activities. As we saw in Chapter 1, "playing the stranger" – the adoption of the anthropologist's stance of close scrutiny of activities the participants take for granted – has become a standard strategy for science studies. The whole genre of laboratory studies, not to mention numerous other investigations of career paths, disciplinary characteristics, and the rituals of academic life, have built upon the ethnographic premise that behavior can be decoded as meaningful action.

One consequence of the popularity of this approach, as has already been noted, is that studies have tended to focus on relatively limited domains of time and space. Ethnographers have set the example by dealing preferentially with institutions or communities that can be observed in a restricted setting over a limited period of time. These are the conditions under which anthropological fieldwork must be accomplished. Thus, the suggestion conveyed by carrying over the term "culture" from anthropology to science studies is that the relevant units of analysis are configurations of a fairly limited size. All cultures, including scientific ones, are "local." Geertz has put the point in a typically provocative way: "Indeed, when we get down to the substance of things, unbemused by covering terms like 'literature,' 'sociology' or 'physics,' most effective academic communities are not that much larger than most peasant villages and just about as ingrown" (1983: 157).

The supposition that, as a culture, a group of physicists can be approached on the same terms as a peasant village undergirds one of the most sustained and imaginative anthropological investigations of contemporary science: Sharon Traweek's portrayal of communities of high-energy physicists, at the Stanford Linear Accelerator in California and the KEK laboratory at Tsukuba in Japan, in her book *Beamtimes and Lifetimes* (1988). Traweek draws significantly upon anthropological concepts and methods. Her footnotes refer extensively to comparable studies of such topics as how work activities in villages and houses are distributed in space, how people represent themselves in interpersonal interactions, how gender identities are constructed, and how divisions of time have symbolic importance. Evidently, these themes informed her observations in the course of her fieldwork among the physicists, whom she identified as forming "communities" like those traditionally studied by anthropology. The physicists are found clustered around a specific site, the spatial distribution of their activities symbolically signifying their relative status within the group. Their work unfolds in temporal cycles, from planning to completion of an experiment, like the cycles of the agricultural year observed in subsistence communities. In these respects and others, Traweek suggests that the local character of her study is not simply a reflection of the limitations of the fieldwork method but represents the inherently restricted range – in time and place – of human activities.

This assumption informs Traweek's extension to the physics community of the postulate Emile Durkheim developed in connection with "primitive" religious beliefs, namely, that "a culture's cosmology – its ideas about space and time and its explanation for the world – is reflected in the domain of social action" (Traweek 1988: 157). Along with an "ecology," a mode of social organization, and a way of patterning their life cycle, the physicists possess a "cosmology" that is a reflection of how their society is ordered. The relationship is a complex one, however, about which Traweek says disappointingly little. She points to an apparent contradiction between the cosmology the physicists have created and the social world they inhabit: Their mental world appears as the polar opposite of the time-bound, contingent, localized, and gender-biased realm of social action. Although their human existence is bound by these constraints, the physicists are said to have constructed "a culture of no culture, which longs passionately for a world without loose ends, without temperament, gender, nationalism, or other sources of disorder – for a world outside human space and time" (162).

How a cosmology like this might have arisen from such a locally bounded culture remains something of a puzzle within the terms of Traweek's analysis. For Durkheim, however, it was the critical problem of the sociological study of religion, and the answer lay in the high degree of generality conferred upon mental constructions by their status as products of society as a whole. Durkheim argued that notions of the transcendent and the eternal lie well beyond the experience of any individual; therefore, they could only be created by the social totality. The objectivity and temporal stability of conceptual thought were, for him, the signs of its origin in collective representations that extend beyond those which individuals are capable of forming alone (Durkheim 1915/1976: 415–447).

In relation to Traweek's study, this raises the question: Is the local culture of physics laboratories the right context for understanding the formation of collective representations among physicists? Or, is the spatially and temporally restricted culture that is apparent to anthropological fieldwork the appropriate one for explaining the origin of physicists' ideas? The physicists themselves would probably say, no. They tend to hold that their concepts derive from direct contact with the transcendent and eternal realm which they describe. Nature is just "like that." Constructivist analyses have not accepted this as an adequate explanation, but neither have they all been limited in their invocation of the social realm to the restricted setting of a small experimental group. Other analysts, including others inspired in part by anthropology, have insisted that the material practice of scientists and the knowledge they produce need to be taken more seriously. Doing so, they are led to suggest that the collective in which scientists operate is considerably more extensive than a small group of investigators at a single laboratory. They have argued, in fact, that in several respects the "cultures" of science are quite different from those which anthropologists have been accustomed to recognize.

Traweek acknowledges this, to the degree that she describes the physicists as being embedded in "networks" that extend well beyond the laboratories in which they currently work. By networks, she means the interpersonal connections through which preprints and gossip are diffused, graduate students are exchanged, and discussion about findings and goals is channeled. But, although larger than single laboratories, the networks remain confined to the elite of high-energy physics, a community that numbers (according to the estimates of its leading members) no more than a thousand people worldwide (Traweek 1988: 3, 106–107). Beyond this, Traweek's networks do not extend. In the light of other constructivist studies, this appears a rather drastic restriction on the scope of the analysis. As I have shown in this book, studies of the extension of experimental knowledge beyond its site of origin have pointed to the importance of the widespread diffusion of technical artifacts, representations, and embodied skills as means of translating scientific knowledge throughout society at large. It has been argued that it is by means of these extended channels of communication that facts can be made to travel away from their points of origin.

These configurations are not only more extensive than the connections between members of a single specialist community; they are also importantly heterogeneous. An understanding of the diffusion of scientific knowledge from the laboratory has been shown to require acknowledgment of the very diverse elements involved in practice at the research site and of its multiform links with the world beyond. Hence, the extended networks invoked by analysts of scientific practice have included many different kinds of entities: trained and untrained personnel, images and representations, discourse and texts, artifacts and instruments, raw materials and engineered organisms. It is by tracing the extension of networks of this kind, which are both pervasive and heterogeneous, that constructivists have sought to map the global spread of science and technology.

Furthermore, as I have also indicated already, other studies have demonstrated how scientific practices alter social relations, the material environment, and even relations of time and space, in the world beyond the site of experimental work. In other words, the identities of cultures are profoundly transformed by scientific and technical practices. Far from appearing as a stable context surrounding experimental work, culture comes to look like something that is disaggregated and newly reassembled by that work. Scientific practice draws upon very diverse elements of the culture in which it is set, and it reconfigures those elements profoundly and extensively. As Andrew Pickering (1995a) has recently argued at length, the practices that produce knowledge and its associated artifacts also remake the culture that surrounds them. That culture cannot therefore be assumed to provide solid ground on which to rest explanations of those practices.

As it happens, these conclusions are quite consistent with the attempts of some contemporary anthropologists to rethink the notion of culture invoked in their discipline. James Clifford has sketched his own critical archaeology of the term "culture." He notes that the anthropological usage of the term, originating in the late nineteenth century, continued to draw upon the aesthetic connotations that Williams traced back to the reaction against the Industrial Revolution (Clifford 1988: 233–235). He remarks that, "Culture, even without a capital *c*, strains toward aesthetic form and autonomy. . . . [It] orders phenomena in ways that privilege the coherent, balanced, and 'authentic' aspects of life" (232). Cultures have thus classically been viewed as local, functionally integrated, and organic entities, changing (if at all) only slowly and continuously. Interactions between the culture being studied and the outside world were regarded as disturbing "noise," to be eliminated as much as possible.

This conception of culture, Clifford notes, has now been undermined by the ending of European colonialism and the growing interconnection of the world's peoples. Changing historical circumstances have suggested that interactions among social groups are not simply extraneous conditions that disrupt the integrity of cultures, but are actually essential to their identification and definition. Anthropologists should be aware, Clifford has argued, of the degree to which their discipline's objectification of particular cultures was a result of the specific trajectories of the journeys that brought them into contact. We should recognize, he suggests, "culture *as* travel," a condition that is shared by anthropologists and many of the people upon whom they report. What has to be acknowledged is the mutually implicated identification of distinct cultures, which is the outcome of each interaction between them (Clifford 1992).

A similar point has been made by Bruno Latour. He locates the roots of the concept of culture in an anthropology that was constituted by expeditions, encounters, and returns to the home base. Classical European anthropology was constructed at the centers of knowledge networks, where traces of groups encountered throughout the world were gathered together. This implies that awareness of different cultures is the result of "crossing other people's path," in other words, of the displacements of travel and the circulation and accumulation of inscriptions (Latour 1987: 201). Thus, the concentration of anthropological knowledge in the centers of European imperialism reflects the imbalance of trade and capital accumulation that favored those locations. The fact that, as Latour remarks, the West does not really regard itself as a culture, but reserves that term for others, is tied closely to the unique history of Europe as a global colonial power.

Both Clifford and Latour draw an important double implication from their recognition that the notion of culture has been tied into the historical patterns of colonialism. They suggest *both* a leveling of the differences that have been thought to distinguish the modern West from the cultures it has observed *and* a new characterization of what it is that separates the two sides. Both writers reject the supposition that European civilization has been given its advantage by some inherent superiority of intellect. Latour strenuously debunks the idea that the "great divide" has a basis in different cognitive abilities. Presumably, neither writer would be comfortable evoking echoes of "manifest destiny," or the special dispensations of Christian providence, to explain European domination. In these respects, they advocate a reduction of the peculiar privilege that has been accorded the West in order to account for its advantage in scientific capital.

Clifford and Latour, however, both respecify what it is that has historically differentiated the knowledge of Western colonial powers from that of other cultures. For Clifford, the crucial activity has been that of collecting, which has been a consistent feature of Western appropriations of non-European cultures (1988: 215–251). He traces the continuity of this enterprise, from the early-modern cabinets of curiosities, which boasted artifacts from Native American, African, and Asian peoples, to the appropriations of "primitive" art by modernist artists of the early twentieth century. For Latour, it is science that has created the perception of a "great divide" between the modern West and other cultures (1993: 91–129); but Latour's science is really collecting under another name, characterized as it is in terms of the accumulation of traces and specimens at a "center of calculation."

The differences between Western scientific knowledge and the cultures it has historically subjugated thus come down to differences in the scale of the displacements in which each was engaged and the length of the networks they were able to sustain. The European advantage in knowledge of other cultures was the result of successful travels to peripheral regions and the transfer of traces of the people living there back to the center. Latour has described this particularly well. Inequalities in the scale of mobilization of resources do, in his view, distinguish Western from non-Western systems of knowledge. However tempted we might be to deconstruct a privileged European tradition of anthropology, we have to acknowledge the asymmetry that created a scientific knowledge of many cultures that had nothing like the same degree of knowledge of one another. Latour diagnoses the difference as simply one of the extent of the networks that the two sides succeeded in building, those centered on Western research institutions having been (at least until quite recently) the largest: "In other words, the differences are sizeable, but they are only of size" (Latour 1993: 108).

There is a dual implication here also for "cultural studies of science." On the one hand, application of the methods of the anthropologists to scientists themselves might be seen as warranted by the moral goal of reducing Western and non-Western cultures to a level of comparison. Why not scrutinize scientists in highly prestigious physics laboratories with the same techniques that have traditionally been used for "primitive" peoples? This has at least the advantage of reducing them to a human scale and reminding us that they are not possessed of any superhuman intellectual abilities. It has the shock value and the refreshing benefits of a "carnivalesque" reversal of the normal hierarchies of status.

On the other hand, however, carnival usually ends in a feeling of let-down and a reassertion of the hierarchy; in the long term, it may even strengthen the prevailing order. Similarly, an approach that "anthropologizes" small groups of scientists appears to reduce science to a culture like any other, but does so in an ultimately unpersuasive way. The task of understanding science and the privileged status it enjoys requires some acknowledgment of the size of the domain over which its activities extend. What is needed is to widen the optic, to examine the knowledge-making practices of the actors and the broader relationships in which they are thereby embedded. In this way, we can try to map the large-scale networks in which practitioners of the sciences are involved, which extend much further than those exploited by the members of many other human cultures. It is only if they cease to be blind to these factors that cultural studies of science will truly contribute to the critical understanding of scientific knowledge in its globally pervasive role.

REGIMES OF CONSTRUCTION

In the previous section, we looked at some of the limitations of certain attempts to study science as a culture. The category "culture," preserving the aesthetic connotations of its historical origin, tends to direct attention toward the homogeneous, autonomous, and strictly localized context of scientific practice. This approach ignores the heterogeneous variety of the resources used in science (texts, artifacts, spoken discourse, materials, images), the varied connections they make with a sometimes very extended milieu, and the capacity of scientific practice itself to fragment and reorder its surrounding culture. For this reason, the problem of the nonlocalized character of scientific knowledge, and the various means by which it is accomplished, tends to be overlooked.

The issue here is, of course, the "problem of construction," introduced in Chapter 1. As we have seen, research pursued within the constructivist framework has tried to tackle the issue of how scientific knowledge, made with resources concentrated in certain localities, can be found valid away from its site of production. Rejecting the response that this is a simple outcome of science's truth to nature, constructivist studies have attended closely to the roles of human agency and material practice in the process. Work on skills and discipline, on instruments and representations, and on discourse and mapping, has all been addressed to this question. A substantial body of empirical research has been devoted to what used to be thought of as purely philosophical problems – those of replication and induction.

Having surveyed many of the component elements, we now have to consider how they have been put together. In this section, we look at studies of the infrastructure of science and technology – the extensive networks that enable scientific facts and artifacts to travel. Several historical studies have been devoted to charting the construction of these networks, in which heterogeneous entities are linked together, and which extend beyond the boundaries of local cultural realms. By considering the role of these extended configurations in the widespread reproduction of phenomena and artifacts, we can substantiate our claim that constructivist sensibilities, and the work they have inspired, offer a valuable complement to studies of science as a local culture.

Let us begin by returning to Bruno Latour and the perspectives he has put forward for tackling the problem of construction. Latour talks about both facts and things as being "black-boxed" for transmission along the "networks" by which science and technology are extended through time and space. These networks do not render laboratory-made knowledge completely universal, but they do sustain its reproduction at numerous sites that are linked with one another. The networks are not infallible; they may break in cases where machines cannot be made to work or factual claims are not repeated unchallenged. But the chances of smooth transmission of laboratory-made knowledge can be increased by modifying the outside world so that it becomes, in crucial respects, more like the laboratory. In this connection, Latour emphasizes the importance of *metrology*, the enterprise that works to secure the compatibility of standards of measurement in different locations. Creation of standard means of measurement is an essential precondition, according to Latour, for the replication of experimental effects at secondary sites; it helps to prepare the "landing strips" that are needed for phenomena to be reproduced, or for instruments to work, away from their place of origin. As he puts it:

> *Metrology* is the name of this gigantic enterprise to make of the outside a world inside which facts and machines can survive. Termites build their obscure galleries with a mixture of mud and their own droppings; scientists build their enlightened networks by giving the outside the same paper form as that of their instruments inside. In both cases the result is the same: they can travel very far without leaving home. (Latour 1987: 251)

For Latour, metrology is an enterprise of enormous scale and expense, the magnitude of which is overlooked because of the common notion that scientific knowledge diffuses by virtue of its inherent truth. Time, for example, is not, in Latour's view, intrinsically universal; rather, "every day it is made slightly more so by the extension of an international network that ties together, through visible and tangible linkages, each of all the reference clocks in the world and then organises secondary and tertiary chains of references all the way to this rather imprecise watch I have on my wrist" (1987: 251). This is, of course, merely to gesture toward the technical and social accomplishments that together have made possible the general availability of standard measures of time.

A fuller historical account would have to include things like the calendar reforms of the ancient and early-modern periods, the manufacture of clocks and watches, the shift from agricultural to industrial patterns of work organization, the spread of metropolitan time by railways and telegraphs, the formalization of worldwide time zones, and the contemporary work of national standards bureaus with their atomic clocks (Landes 1983; Thompson 1967; Kern 1983: chap. 1). Similar accounts could be given of the standardization of units of length and weight measurement, in which the creation and adoption of the metric system would assume an important place (Alder 1995).

Latour, however, extends the meaning of metrology beyond its normal application to the fundamental physical constants. He applies the term more broadly to all the activities by which the material and human conditions for the production of knowledge are standardized. Metrology, in this wider sense, includes such projects as the manufacture and distribution of purified samples of chemical substances, the breeding of special strains of plants and animals for experimental purposes, the circulation of standard printed forms for reporting information, and the training of personnel in procedures for calibrating instruments. All of these activities form part of the massive organizational enterprise by which the world is made over in the image of the laboratory, and hence is made receptive for laboratory-made knowledge. By recording these activities, we come to appreciate how, as Theodore Porter has written, "What we call the uniformity of nature is in practice a triumph of human organization – of regulation, education, manufacturing, and method" (Porter 1995: 32).

A number of historians have risen to the challenge of analyzing the development of the metrological enterprise and delineating its relations with the laboratory science that it has made so powerful. Schaffer (1992) has explored the role of Maxwell and the Cavendish laboratory in the standardization of units of electrical resistance, a project of critical importance for the extension of telegraph networks in the mid-nineteenth century. When he was recruited to Cambridge in 1871, Maxwell brought with him a calibration instrument developed in conjunction with the engineer Fleeming Jenkin in London, in work coordinated by the standards committee of the British Association for the Advancement of Science. The apparatus comprised a coil of wire spinning around its vertical axis in the earth's magnetic field, with its motion regulated by a very precise gear and governor mechanism. A galvanometer connected to the wire and mounted at its center would produce a constant deflection, which would depend upon the coil's diameter, its rate of spin, and its resistance. By measuring the angle, speed, and diameter, the resistance of the wire coil could thus be calculated (Schaffer 1992: 27). At the Cavendish, Maxwell institutionalized the regime in which this apparatus was used to calibrate coils that served as resistance standards, being boxed and distributed to engineers who used them to locate faults in telegraph cables. Maxwell and his successor, Lord Rayleigh, mobilized material, literary, and social resources to connect the standards calibrated in the Cambridge laboratory with the measurements of telegraph engineers working on the Atlantic cables or across the far-flung British Empire. The Cavendish physicists, in Schaffer's words:

> told their pupils and customers where to buy the robust instruments, where to learn the right techniques, and what morals would be appropriate for laboratory workers. Then they stated that the absolute system depended on no particular instrument, or technique, or institution. This helps account for metrology's power. Metrology involves work which sets up values and then makes their origin invisible. (1992: 42)

Joseph O'Connell (1993) has taken this story further with an account of the formulation of international standards for electrical units in the late nineteenth century. He reiterates the importance of metrology in the creation of the infrastructure that supports the international circulation of modern science and technology: "The visible circulation of TV sets and computers from one country to another . . . is only possible because of the invisible circulation of standards to all the factories at which computers and TV sets are made, to all the transmitting stations which broadcast TV signals, and to all the utility companies which produce electrical power at a set voltage" (O'Connell 1993: 164). The picture is, however, a complex one, because various strategies may be adopted to construct metrological networks. O'Connell describes some of the alternatives, indicating that different political models were relevant to the different systems produced.

Thus, in the 1860s, the approach endorsed by a committee of the British Association for the Advancement of Science was as follows: The unit of resistance (the "ohm") was *defined* in absolute mechanical units as ten million meters per second. That unit then had to be *realized* in practice in Maxwell and Jenkin's spinning coil experiments in the mid-1860s, which produced a coil that could serve as the *representation* of the standard. Precautions had to be taken to maintain the accuracy of the coil against its expected degeneration over time. This was done by creating a "parliament" of five pairs of coils, each pair made of a different metal or alloy. By talking of a "parliament," the committee invoked a notion of majority voting: If one coil or a pair degenerated from the original standard resistance, it was assumed that this would be revealed by its divergence from the value sustained by the others. The ten standard coils were deposited in the Kew Observatory, where they were kept for reference and used to prepare resistance boxes for circulation throughout the empire (O'Connell 1993: 137–143).

O'Connell contrasts this procedure with the contemporary German ap-

proach, inspired by Werner Siemens. Siemens proposed an arbitrary definition of a unit of resistance (which came to be named after him), as
that of a one-meter-high column of mercury, of one square millimeter
cross-section, at zero degrees centigrade. The definition was arbitrary,
but Siemens's publication of his methods of calibration enabled it to be
reproduced by anyone with sufficient resources of skill and equipment.
Siemens claimed a significantly greater degree of accuracy in the reproduction of this standard than could be achieved by the British methods.

At the International Electrical Congress in Paris in 1881, a compromise
between these two metrological procedures was reached. The British
"ohm" was accepted as the unit of resistance, and its absolute definition
proclaimed. The definition was to be realized by the spinning coil apparatus as used at the Cavendish. But this definition was to be represented by a mercury column (of 1 mm^2 cross-section at 0°C), of a length
to be determined to correspond to the absolute unit. Thus, the "legal
ohm" came to be specified as the resistance of a mercury column of 106
cm, and the publication of this specification was expected to make the
standard available to any laboratory that wished to replicate it. In effect,
the German procedure had been made the basis of the legal unit, but
this had been grafted onto the British method of absolute definition.

As O'Connell shows, however, the international agreement did not in
fact produce uniformity in metrological practices. National variations
persisted. France and Germany adopted the ohm and its mercury-
column representation, but ignored its supposed absolute definition. In
Britain, on the other hand, the representation of the unit in a mercury
column was a dead letter: the standard was identified with a platinum
coil maintained in the Board of Trade laboratory in London (1993: 141–
145).

O'Connell's analysis makes an important general point: Metrology is
an inherently social and political process. National rivalries evidently
shaped the choices of metrological procedures, and this continued even
after international agreement had ostensibly been reached. Schaffer invokes further elements of the political context in his overlapping account.
Maxwell's work at the Cavendish was represented as part of the Victorian enterprise of work discipline and moral surveillance, and as essential for the telegraphic links that sustained the integrity of the British
Empire. These contextual factors are by no means simply "external" to
the actual practices of construction of metrological networks. O'Connell
shows, on the contrary, that political values may be entrenched quite
fundamentally in decisions made about the structure of those networks.
The model of a "parliament" of standard coils, used to calibrate sets of
coils that are then circulated worldwide, comprises a quite different metrological polity than the German method of publishing a definition of
the standard.

O'Connell reinforces the point with a discussion of the shift in recent decades to "intrinsic standards" of electrical and other measurement units. These are "physics experiments that other laboratories can perform to create the volt, second, ohm, or various temperature points right on their premises to the highest accuracy that is legally recognized" (1993: 152–153). By exploiting recently discovered microphysical phenomena, laboratories can reproduce measurement standards with as much accuracy as any national standards institution, provided they have the necessary resources of equipment and expertise. The effect has been to dispense with the need periodically to call in approved instruments, such as standard voltage cells, that embody the units in a form certified by the central institution. Regular visitations by certified standard instruments are no longer necessary to calibrate laboratory apparatus. As O'Connell puts it, the "Catholic" routine of "periodical sacramental redemption from error" has been replaced by a kind of "Calvinist reformation in metrology" (154).

The religious labels indicate very well the quite different structures that metrological networks can assume. The development of intrinsic standards has accompanied a shift toward reliance on the *word* (the specifications for reproducing the standards), as in the Protestant Reformation, rather than on the "sacraments" (visitations from certified standard instruments). But this, of course, is not the whole story. Intrinsic standards are only reproducible because of many other metrological networks that O'Connell does not discuss: those responsible for circulating and sustaining the equipment, materials, and skills needed to reproduce them at many different sites. Nonetheless, the general point is that quite different kinds of networks may be built and maintained, involving different emphases on circulation of instruments, written instructions, or tacit skills, or entrenching those elements in different locations in the system.

This has general implications for Latour's picture of the networks that allow scientific knowledge to travel beyond its site of origin. Latour portrays metrological networks as ramifying from single sources, which he frequently identifies with the laboratory of a prepotent individual, such as Louis Pasteur (Latour 1988a). The same privileged sites can serve as "centers of calculation," exerting effective domination over the peripheries of the network by capitalizing upon accumulated "immutable mobiles." As Shapin has recently noted, this account amounts to a "descriptive vocabulary of power." Notwithstanding Latour's disavowal of any knowledge of individual motivations, Shapin remarks that "the agent deploying these resources is recognizable from Machiavellian and Hobbesian accounts of human nature: Pasteur is displayed as animated by a will to power and domination, and his readers' decisions to acquiesce or submit are treated as those of pragmatic maximizers-of-marginal-advantage" (Shapin 1995: 309).

The more historically nuanced accounts of Schaffer and O'Connell suggest the need for a more flexible model. They point to variations in the constitution of networks, involving different patterns of distribution of materials, artifacts, human resources, and discursive specifications. They recognize the possibility of multiple and competing centers of power. And they raise the issue of the relations between the networks of "technoscience" and what are more traditionally recognized as political structures – local, national, international, and imperial. Different polities or regimes of construction, in other words, might well be uncovered by a more sensitive historical approach.

This aspiration can draw support from Thomas Hughes's magisterial treatment of the construction of electrical power supply systems in Europe and North America, in the late nineteenth and early twentieth centuries, *Networks of Power* (1983). Latour invokes Hughes's book as a case study of network building, since it charts the extension across space and time of the electrical supply systems that have provided the infrastructure for much subsequent scientific and technical innovation. Hughes maps the spread of the systems, treating them as "coherent structures comprised of interacting, interconnected components," which, he notes, "embody the physical, intellectual, and symbolic resources of the society that constructs them" (Hughes 1983: ix, 2). He exposes the interaction of technical and social constraints, which demanded solutions that simultaneously addressed both kinds of conditions. The elements of each system – Thomas Edison's incandescent filament bulb, for example – had to satisfy the multiple constraints of the network in which they were to function. In this case, the commercial price of the copper used in the conducting cables dictated Edison's search for a high-resistance filament that would not demand a high supply current (31–34).

Hughes is admirably sensitive to the commercial and political landscapes in which the electrical supply networks were constructed. In the case of the failure of Edison's licensed company in London in the 1880s, he notes the need to provide "other than strictly technical explanations" (1983: 56). In the English capital, the legislative regime, which mandated public ownership of utilities companies after twenty-one years of private operation, is seen as the crucial obstacle to entrepreneurial development. Hughes proposes that local political conditions continued to obstruct the expansion of electrical supply in London through the 1900s and 1910s. With examples of this kind, he displays the intimate connections between electrical supply systems and the social and political setting in which they were constructed and which they also significantly reshaped.

On one level, Latour is correct in perceiving Hughes's account as a case study of network building in the sense that he defines it. Electrical supply systems are constructed from the center toward the periphery; they do suggest a model of the prepotency of the center, which brings elements of the surroundings progressively within its domain. Hughes points to Edison's own conception of his whole system as a single "machine," designed from a central vantage-point so that all its parts would function harmoniously together (1983: 22). Even while he shows the system designers negotiating with the elements of their surroundings, Hughes's vocabulary of "evolution" of systems suggests a model of natural growth from the point of origin. His notion of "reverse salients" – adopted from military usage to designate sites where the advancing front is hindered, and which become the focus of intensified efforts to locate and remove the obstacle – also suggests a view from the center and an assumption that outward expansion could ideally proceed without any efficient cause.

Hughes's study also, however, provides resources for developing an alternative and more nuanced view of the regimes of construction involved in the electrical power supply story. Not all of his account is concerned with networks that grew steadily to absorb more of their surroundings. He also describes the intense "battle of the systems" that pitted direct current (d.c.) against alternating current (a.c.) supply schemes in the 1880s and 1890s (1983: 106). Each system can be said to have experienced its own "reverse salients." Some of them were overcome (for example by improved designs of d.c. motors for many types of applications), and some not (for example, the continuing problem of economical transmission of d.c. supply over distances of more than a few miles). But the problems faced by each system were effectively *defined* by the situation of conflict between them; they would not have been identified in the same way if an alternative system had not existed. For instance, fears that a.c. supply was potentially more dangerous were deliberately amplified by supporters of the d.c. system, who staged public electrocutions of animals and employed the rival current for an execution in the New York State electric chair (108). In this case, the environment into which each network sought to expand was not a neutral "flatland"; rather, it was shaped by those who were struggling to build a rival network competing for the same territory.

In this situation, the final organization was accomplished by a consolidation of the two systems. Hughes traces the rise of the concept of a single "universal supply system" and its relation to the consolidation of supply companies in the 1890s (1983: 122–123). The general adoption of the a.c. system was accompanied by such technical innovations as the rotary converter, which enabled assets invested in d.c. equipment to be preserved during the transition. At this point, common standards emerged, not from the extension of a single metrological network, but as a result of pragmatic compromise between competing systems. A choice was made from among a number of possible frequencies for the supply of alternating current. As Hughes puts it, "a general agreement about frequency did not come through the establishment of one frequency's obvious technical superiority over the others; rather, a spirit of flexibility and compromise among the various utility interests, and especially among the manufacturers, was primarily responsible for the agreement" (127).

This outcome reminds us of the compromises O'Connell records in the creation of standards for units of measurement. Evidently, common standards are not always the result of networks expanding steadily from the center into a neutral surrounding terrain, as Latour's picture might suggest. When networks are in competition – when terrain is being contested between them – compromise may be required. And this agreement upon common standards, whether of manufactured products or of measurement units, can enable a mutually profitable exchange to take the place of destructive rivalry. The processes of agreement and exchange, and the formulation of common specifications or units to govern such trade, deserve more emphasis than they are given in Latour's "agonistic" model of how science and technology have achieved their global extension. Metrology may be at least partially the result of agreement and exchange rather than solely the product of empire building by Machiavellian individuals.

This point can be reinforced by considering what might seem to be an entirely different means for the global extension of scientific knowledge. Theodore Porter has recently completed a study of the role of quantified information in the modern social sciences. Here, we are dealing with "networks" considerably less concrete than those for electrical power supply. Nonetheless, Porter presents his analysis as having implications for the general issues we have been discussing. He introduces his book, *Trust in Numbers* (1995), by referring to studies of the replication of experimental phenomena, noting constructivist research on the local specificity of laboratory cultures, the need to transfer tacit skills, and the role of standardization and metrology in facilitating widespread replication.

With these themes in the background, Porter moves on to consider the role of quantification in the construction of "objective" knowledge about society. "Objectivity" is to be understood as a moral and political ideal, referring to the "rule of law, not of men. It implies the subordination of personal interests and prejudices to public standards" (Porter 1995: 74). What is taken for objective knowledge is information that is readily accepted by different individuals and groups because it appears untainted by the particular circumstances of its origin. Quantified information – social or economic statistics about the aggregated properties of persons or things – possesses this property of readily moving across cultural boundaries. Figures, stripped of the interpretive work that goes into compiling them, are very mobile signifiers. As Porter puts it, "The remarkable ability of numbers and calculations to defy disciplinary and even cultural boundaries and link academic to political discourse owes much to this ability to bypass deep issues" (86).

Porter goes on to suggest how the prevalence of quantified data in the social sciences, and in the political realm that intersects with them, is not the outcome of an expanding and monolithic scientific enterprise but of insecure and contested disciplines operating in a milieu of considerable cultural diversity. Particularly in the contemporary United States, quantified data are prized and legalistic administrative procedures routinely employed to display openness and fairness. The rule of law and the quantified social sciences provide means of consensual governance in a society that lacks deeply rooted traditions of social authority. Techniques such as standardized educational testing, cost-benefit analysis, or opinion polling become popular because numerical objectivity commands the consent of a culturally diverse population. Quantification in the social sciences is, on this reading, a testimony to the *lack* of trust, on the part of society in general, in the authority of the experts.

In its general implications, as Porter spells out, this picture casts doubt on the vision of scientific practitioners as an autonomous community with a steadily extending influence over society at large. Although small-scale communities of specialists might persist in certain fields, such as high-energy physics, the prevalence of quantified methods in the social sciences is a sign that groups of practitioners in those disciplines are weaker, more loosely bounded, and more extended – more like *Gesellschaften* ("societies") than *Gemeinschaften* ("communities"), to use the terms introduced by Ferdinand Tönnies. Weaker disciplines, or those under more pressure to answer to outside bodies, require more explicit and objective procedures in order to protect their practitioners from imputations of impropriety or bias. The outcome, of course, is that they restrict their claims to results that can be reported quantitatively. Porter concludes: "Some of the most distinctive and typical features of scientific discourse reflect this weakness of community" (1995: 230).

In Porter's view, then, the application of quantified measures to many realms of human affairs reflects the operations of technical expertise in a diverse and contested political landscape. Rather than steadily and progressively expanding their reach over a neutral terrain, the number crunchers in the social sciences have gained influence by offering a "lowest-common-denominator" language in which different groups can communicate about the state of society. Quantified data are tokens of exchange between factions that both compete and cooperate; they are not the tools of a powerful monopoly of experts steadily expanding their control over the social domain. Numerical methods and results might be said to have provided a "pidgin" language for social intercourse between different communities. It is for this reason that the quantified social sciences have achieved their remarkably widespread acceptance, notwith-standing the suspicion and condescension with which they are still regarded in some quarters.

As an alternative to Latour's model of networks built from a prepotent center, Porter offers a rather different view of the extension of scientific knowledge. To decide which model is applicable to any particular case would require careful attention to the specific details of the situation: the kind of knowledge in question and the context in which it is being de-ployed. The medium by which information is carried is clearly of par-ticular importance. Numbers can be mobilized very rapidly and easily, by a range of techniques from tallies on clay tablets to computers. Rather different technologies undergird the translation of words and visual im-ages, making possible in each case quite distinct patterns of distribution. Each new mode of communication – the printing press, the telephone, photography, computer networks – makes possible a new geography and economy of knowledge.

When scientific knowledge has been closely tied to material artifacts, networks might reasonably be seen as constructed outward from the centers of manufacture and distribution. Many of Latour's examples of networks expanding from the center draw upon the histories of machines made by industrial mass production: computers, the diesel engine, the Kodak camera, and so forth. Others refer to the historical era of European global dominance. Much of the history of science and technology over the last two centuries has been entwined with the growth of centralized manufacture following the Industrial Revolution, and with the coinci-dent European colonization of other parts of the world. But, with the breakdown of the pattern of industrial production and international re-lations that characterized that era, we have recently become alert to the possibilities of alternative distributions of authority and of knowledge. One aspect of this is the "postcolonial" research in anthropology, history, and cultural studies, which, while not disputing the inequalities of power that structured relations between the imperial center and the colonial peripheries, has indicated the surprisingly subtle ways in which those inequalities were negotiated locally (cf. Prakash 1992). Along with this research, as we saw in the first section of this chapter, has come a read-iness to reconsider the dimensions of cultural formations, and even the definition of culture. In the postcolonial world, the distinctive patterns of material distribution and cultural exchange that characterized the era of European domination have come to seem temporary and less than completely global.

Other recent research has focused upon the ways in which, prior to the triumph of industrial mass production, the human body provided a readily available – but problematic – means of transmitting natural knowledge. In the eighteenth century, as social historians have noted, displays of the human body were routinely employed for the assertion of various kinds of power: monarchical, aristocratic, judicial, bourgeois, and plebeian. Schaffer has considered how the body was exploited in this period to communicate knowledge of natural phenomena such as static electricity or "animal magnetism" (1994b). Drawing upon sociologists' studies of the role of embodied skills in the mobilization of experimental phenomena, Schaffer considers how bodies could themselves become "instruments" for this purpose. The telling cases are those in which the tool failed to function, in which the body lost its transparency as an instrument and became the focus of attention itself; for example, when what purported to be instances of animal magnetism were diagnosed as delusions of the imagination. Such problematic episodes highlight, by contrast, the normal functioning of the body as a convenient instrument for the public construction of experimental knowledge in the eighteenth century. In the revolutionary upheavals at the end of the century, Schaffer suggests, use of the body as an experimental instrument underwent drastic reappraisal. On the one hand, the scientific practitioner came to be viewed as a "genius," distinguished by a degree of detachment of the mind from the trammels of the body; on the other hand, phenomena were increasingly recorded by self-registering machines, the products of industrial manufacture. Thereafter, replication of experimental effects was to be much more closely tied to the circulation of manufactured mechanical devices.

There remains, nonetheless, much scientific knowledge that continues to be focused upon the human body. For such sciences as medicine, psychology, and physical anthropology, the body serves both as an object of study and as a means of mobilizing the knowledge created. The body is shaped and interpreted by the work of practitioners of these sciences, but it is then exploited as a resource for making their findings reproducible elsewhere. This duality of the body – as object of study and instrumental resource – has been emphasized above all by studies of the role of gender in science. Feminist scholars have very plausibly insisted that the cultural fabric of "gender" is not biologically determined: The whole apparatus of what is culturally designated male or female cannot be reduced to the biological fact of the existence of two sexes. But the same scholars have greatly deepened our understanding of how the biologically reproduced human body provides the substratum on which cultural constructions of gender are built and sustained. Even while they decode gender as a product of culture, not biology, feminist analyses have demonstrated the pervasive use made of biologically reproduced features of humans for the creation of cultural formations of identity, domination, and subversion. Hence, as Evelyn Fox Keller has pointed out (1989), gender analysis has tended to destabilize the polarity between nature and culture at the same time as it has exposed the cultural roots of what was previously taken to be natural.

This aspect of feminist scholarship has a significant bearing upon our understanding of the role of human beings in technological and scientific networks. The human body and mind have been treated both as (natural) objects of investigation and as (cultural) instruments of scientific practice. Yet, the boundaries between nature and culture, and object and instrument, tend to be quite permeable, and the polarities are thus liable to be unstable. Humans frequently refuse to behave as docile instruments, declining to play the part allotted to them, for example, by doctors or psychiatrists who want to establish the replicability of medical or mental phenomena. They have proved much more difficult to confine, breed, and engineer than fruit flies.

Mary Poovey has given an example of this in her study of the role of the female body in medical writings about the use of chloroform in nineteenth-century obstetrics (1987). Victorian doctors, arguing about the appropriateness of anesthetizing a mother in childbirth, denied a voice to their female patient and filled the space of her silence with their own interpretations. Some asserted the inherent modesty of the female patient, who would deplore the erotic feelings and gestures that chloroform sometimes elicited. Others argued that the chloroform revealed – in precisely these effects – a natural female immodesty. Some doctors claimed that women were naturally healthy; others that they had an inherent disposition to pathology. The silence of the female patient herself, however, permitted continuing ambiguity and indeterminacy, due to what Poovey calls the "metaphorical promiscuity of the female body" (1987: 153). The female body, in Poovey's reading, retained the character of a threat to stable scientific representation and control. Poovey concludes: "As a man-made technology, . . . anesthesia did not control the body so much as it disclosed its problematic capacity to produce meanings in excess of what the 'exhibitors' of the technology intended" (155–156).

In talking of the body's "capacity to produce meanings" that go beyond medical discourse, Poovey might be said to have herself filled the space of a pregnant silence by postulating an ontological role for the body. She asserts that the body is capable of disrupting and undermining its representations; yet she does so solely on the evidence of conflicts and equivocations registered within medical discourse itself. To assert that the body can escape from representational mastery in this way goes beyond what the discursive evidence can warrant. Nonetheless, the tensions and uncertainties within medical discourse, which Poovey perceptively isolates, do have to be acknowledged. At the least they signify the troubles that tend to surround use of the body as object and instrument of scientific knowledge.

While they can provide flexible and pervasive means for extending the sway of natural knowledge, human beings can also be remarkably problematic and ambivalent resources. The body offers its biologically reproduced features for the replication of scientific knowledge and effects, but it is disciplined only with difficulty and always incompletely. Humans remain very "soft" instruments for translating natural knowledge. This is bound to complicate the story of the construction of those technoscientific networks in which they are implicated. And they remain implicated in many places, however much embodied skills are displaced and redistributed by the circulation of the "hard" (manufactured) instruments of modern science and technology. In some respects, in fact, human bodies are becoming ever more intimately integrated within systems composed of machines and other organisms, as Donna Haraway's notion of "cyborgs" recognizes (1991). The changes Haraway points to – including the development of new information- and biotechnologies, and the coincident rise of the politics of gender and sexuality – oblige us to question again the respective roles of humans and machines in the construction of modern technoscience as a global system. This poses a challenge to historians as much as to anyone else concerned with understanding the modern world.

Coda: The Obligations of Narrative

How does one relate a piece of research work? How does one retrieve an idée fixe, a constant obsession? How does one re-create a thought centered on a tiny fragment of the universe, on a "system" one turns over and over to view from every angle? How, above all, does one recapture the sense of a maze with no way out, the incessant quest for a solution, without referring to what later proved to be *the* solution in all its dazzling obviousness[?] Of that life of worry and agitation there lingers most often only a cold, sad story, a sequence of results carefully organized to make logical what was scarcely so at the time.

François Jacob, The Statue Within (1988: 274)

Historians are, by occupation, storytellers. More consistently than anthropologists or sociologists, they practice a discipline traditionally articulated in narrative form. It is a cliché to note that the word "history" is ambiguous, naming both the events of the past and their recounting in discourse. But the ambiguity may be a significant one, with its suggestion that the "storied" character of the discourse reflects a fundamental dimension of human experience itself. The French philosopher Paul Ricoeur has reflected upon this; he proposes that "this ambiguity seems to hide more than a mere coincidence or a deplorable confusion. Our languages most probably preserve (and indicate) by means of this overdetermination of words . . . a certain mutual belonging between the act of narrating (or writing) history and the fact of being in history, between *doing* history and *being* historical" (Ricoeur 1981: 288).

Ricoeur has drawn out the philosophical implications of the connection between the narrative form of history writing and the human experience of living in time. He suggests that, "the form of life to which narrative discourse belongs is our historical condition itself" (1981: 288). Framed by this existential condition, historical narrative possesses a dual reference. Literally, it denotes the events described; figuratively, it represents the human experience in which such events function as parts of meaningful stories. To tell and to follow such stories is the way in which humans understand their experience of time as something more than

186

simple "seriality" (the proverbial "one damned thing after another") and as embedded in processes that are directed toward some meaningful end. As Hayden White has put it, summarizing Ricoeur's view, "What the historical narrative literally asserts about the specific events is that they really happened, and what it figuratively suggests is that the whole sequence of events that really happened has the order and significance of well-made stories" (White 1987: 177).

In the English-speaking world, White's own work has made a substantial contribution to the understanding of narrative in historical writing. He has communicated something of the rich resources yielded by European philosophy for consideration of this issue. Along with Ricoeur, he has brought to bear the work of other hermeneutical philosophers like Hans-Georg Gadamer, semioticians like Roland Barthes, and the "poststructuralists" Michel Foucault and Jacques Derrida. This aside, White has also analyzed at length, in his monumental *Metahistory* (1973), the function of narrative in the writings of the great European historians of the nineteenth century. He has argued consistently against the attempts of purportedly "scientific" historians to eliminate the narrative element of historical discourse, insisting that it remains unavoidably present and fulfills (at least in part) the demand that history have some moral meaning. The moral imperative cannot be discharged simply by recording a series of events, nor by quantitative analysis; it calls for the historian to signal that those events constitute a story of a particular kind. That is done by giving the recorded events the form of a *plot*. To quote White:

> The demand for closure in the historical story is a demand, I suggest, for moral meaning, a demand that a sequence of real events be assessed as to their significance as elements of a moral drama.... When the reader recognizes the story being told in a historical narrative as a specific kind of story – for example, as an epic, romance, tragedy, comedy, or farce – he can be said to have comprehended the meaning produced by the discourse. This comprehension is nothing other than the recognition of the form of the narrative. (White 1987: 21, 43)

The issue of narrative, with its connection to the moral meaning of historical discourse, is an important one to consider in the light of constructivist approaches to the history of science. As we saw in the introduction to this book, the classical tradition of large-scale histories of the sciences, inaugurated in the period of Priestley and Whewell, exploited a progressive narrative form in the service of philosophical claims that were both epistemological and moral. The story of the steady onward progress of knowledge conveyed a lesson about how knowledge was constituted in the mind, and about the conditions of personal character and social setting that were conducive to this. In the wake of the strong challenges recently mounted against classical epistemology, the traditional narrative forms of the history of science now seem unsustainable. But this raises the question as to how the narrative obligation – with its due burden of moral meaning – is to be discharged by historians today. What kinds of stories ought we to be telling?

The question is an urgent one, not least because of concern about the relations between the academic discipline of history of science and its audiences: students, scientists, and the general public. The practical pedagogical duties of historians, their dealings with scientific colleagues in educational institutions, and their involvement with forms of "public history" (such as museum exhibits or television programs), all bring to the fore the disparities between the stories they are inclined to tell and those their audiences are expecting to hear. The recent attacks on radical tendencies within science studies by self-styled defenders of the epistemic credentials of science are surely symptomatic (among other things) of a disappointment that the good old stories are no longer being told.

Reflecting this concern, a recent symposium in London pondered the fate of "big-picture" histories of science. The premise of the gathering was that recent trends in the academic discipline have resulted in a diminution of the scope of historical writing, which has largely abandoned the aim of telling a story of science's universal progress. In this situation, students and others who look for big-picture surveys of the whole subject are liable to be disappointed. John Christie suggested, however, that what is felt to be missing here is not so much spatiotemporal scale as a particular narrative form. He pointed to a lingering nostalgia for the kind of historical narratives that took their rise in the century of the Enlightenment, in the works of such philosophers as Jean D'Alembert, the Marquis de Condorcet, and Adam Smith. For these writers (and those like Priestley and Whewell who followed their lead), science was the appropriate field in which to tell a story driven by a vision of human epistemological capabilities and their temporal realization. This vision generated a mode of emplotment within which events were to be narrated. Human knowledge was to be shown to have advanced along a linear, rising path, its direction always clear even when its forward movement was temporarily obstructed (Christie 1993).

Christie traces the persistence of this narrative mode into the middle of the twentieth century. Notwithstanding the evident differences between the various grounding philosophies – positivist, Kantian, Hegelian, Marxist – he claims that they all constructed narrative unity out of a series of actions construed as homogeneous mental processes (though Marxists added to this a simultaneous and correlated series of material processes). Acts of mental perception and judgment were described in a sequence that unfolded according to its own inherent logic and was consistently oriented toward the present. It was in this sense, he claims, that "big picture historiography could be held to be philosophically plotted" (1993: 397).

Christie's analysis is shared in its broad outlines by Joseph Rouse (1991), who also ascribes the tenacity of certain story lines to the continuing hold of a particular philosophical outlook. Rouse suggests that almost all current philosophies of science (positivist, "post-empiricist," and realist) endorse a single narrative model for its history, namely, inscription within a more extensive drama of the progress of modernity. These "narratives of modernity" integrate the development of science within "stories of progress, which portray modernization as the advance of knowledge, the establishment of human rights and humane practice, political democratization, the creation of wealth, and the control of nature, all to the benefit of humanity" (1991: 147). The prevalence of this narrative mode even creates a space for the development of "antimodernist" narratives, which describe the increased domination of mankind by instrumental rationality, the destruction of the environment, bureaucratized inhumanity, the death of the human spirit, and so on. Such laments, on Rouse's reading, do no more than reiterate the outlines of the narratives of modernity, attaching a reversed valorization to the crucial elements in the tale.

Intriguingly, then, Christie and Rouse give similar narratives of the history of historical narrative; but they differ in their diagnoses of the current situation. For Rouse, almost all contemporary thinking about science remains within the framework of the "narratives of modernity," although he discerns signs of a possible breakthrough to a "postmodern" narrative model, of which certain features can be glimpsed already. Christie, on the other hand, rejects the applicability of the modern/postmodern dichotomy to the current situation, suggesting that what characterizes contemporary historiography is not the eclipse of all "master narratives" (trumpeted by Jean-François Lyotard, the prophet of postmodernism) but the coalescence of interest around a new vision of science stimulated by post-positivist philosophy and constructivist sociology. This new vision suggests the possibility of the reemergence of big-picture narratives, albeit of a rather different kind from the traditional stories of progressive epistemological accomplishment. Christie points toward histories in which the time-embodied nature of scientific practice would be traced and the extension of its material control related to the waxing and waning of large-scale formations of political and social power. Such a history might be written, he suggests, by enlisting as resources a range of conceptual and historical studies of science, including those of Foucault, Latour, and (ironically) Joseph Rouse.

For Rouse himself, however, it seems clear that such a historical narrative would not succeed in transcending the bounds of the discourse of modernity. Indeed, he reads constructivist sociology of science as inscribed within that discourse, albeit somewhat ambivalently. For Rouse, the claims by sociologists of scientific knowledge to an empirical basis for their studies disclose their reliance on traditional epistemology: "Their history of science is also a history of Enlightenment, in which sociologically naive scientists and philosophers of science replace tyrants and priests as the purveyors of superstition to be overcome" (Rouse 1991: 151). On the other hand, the sociologists deny science the global legitimacy it claims, thereby implicitly (according to Rouse) conceding that it *needs* such legitimation to undergird its acceptance. They counterpoise "social" to "cognitive" or "rational" causes of beliefs in a way that continues to play off a dichotomy between "science" and "society" that they claim to have transcended.

These criticisms may be warranted by some of the rhetoric of the early constructivist sociologists, though Rouse provides no citations to support them. Some of the case studies by Collins and others did indeed operate with a kind of dichotomy between "social" and "rational" causes. Collins tended to argue that social causes should be invoked for the adoption of beliefs because the decisions involved could not be entirely explained on the grounds of scientific method. But others carefully disavowed such a polarity. David Bloor's "symmetry thesis" was directed at uncovering the social causes that were always involved in (though not exclusively responsible for) the adoption of beliefs, regardless of our assessment of their rationality. Neither social nor rational causes, in other words, were seen as exclusive of the other. Bloor was simply asking for social causes to be given their due in all cases. He was arguing for transcendence of the dichotomy rather than its reversal and thus, understandably, repeatedly disclaimed any intention of debunking scientific claims to knowledge.

Rouse sees constructivism as unable to move beyond a polarization of the rational and the social that is characteristic of modernity, and hence as remaining hampered by a narrative of scientific enlightenment. But to characterize constructivism as a whole in such a way would be to miss the movement of many of its practitioners toward a more complex understanding of the involvement of their own enterprise in the practices of science. A more detailed and discriminating reading of constructivism would have to acknowledge, for example, Shapin and Schaffer's account of how boundaries were erected between the realms of natural and social knowledge in the course of the dispute between Boyle and Hobbes. As Latour (1990, 1993) has emphasized, the dichotomy can scarcely be said to remain unproblematic as one follows through this analysis. The "reflexivist" move, whereby a critique of scientific practices of representation and argumentation is brought to bear on sociological discourse itself, is present in many varieties in constructivist work, as the dizzying survey by Malcolm Ashmore (1989) has shown. Ignoring these moves,

Rouse has neglected a wide range of developments arising directly from constructivism that lead in the very direction he wants a "postmodern" account of science to go.

Leaving aside the semantically vexed question of whether contemporary science studies are appropriately characterized as "modern" or "postmodern," it is certainly arguable that they bring seriously into question the likelihood that big-picture historiography will return, even in the form that Christie envisions. Rouse is surely right to point to the loss of innocence involved in the constructivist critique of narratives of scientific enlightenment, and this raises problems for the history of science both in relation to its subject matter and in relation to its own methods. At the very least, contemporary historical writing must be framed by some awareness of these problems. In a generally sober and measured account of the implications of current philosophical debates for historical practice, Joyce Appleby, Lynn Hunt, and Margaret Jacob have acknowledged that, "In the nineteenth-century sense, there is no scientific history, nor is there even scientific science"; though these authors nonetheless seek to recuperate a concept of "truth" that will continue to be serviceable to historians (1994: 194). However these problems are tackled, it is apparent that historians have crossed a chasm that divides them from the narratives of scientific development that scientists themselves have typically told. A focus on narrative may allow us to map the extent of this divide, in terms that do not require us to venture into the morass of debate on how to discriminate the modern from the postmodern.

In other recent work, Rouse (1990) has attempted to apply to science itself the category of narrative, claiming not simply that it characterizes a style of writing but that it constitutes "the temporal organization of the understanding of practical activity." In their day-to-day work to impute meanings to their own and others' actions, practising scientists are constantly engaged in storytelling, interpreting the direction of development of past work and projecting it into the future. "Scientific research is a social practice, whereby researchers structure the narrative context in which past work is interpreted and significant possibilities for further work are projected" (1990: 179). Events are only given meaning by their incorporation in such narratives, which individuals are constantly competing with one another to complete. This narrativity of practice is a consequence of the embeddedness of all meaningful action in sequences of events understood as stories. "To act," Rouse proposes, "is to be ahead of oneself, that is, to have some understanding of what it would be to have done the action in question." All actions are "being enacted toward the fulfillment of a projected retrospection, but one which is constantly open to revision, as befits a story not yet completed" (1990: 183–184).

This general account of the narrativity of practice is intriguing, but it does not give any very specific help with analyzing how science is done. Rouse proposes that scientists maneuver their factual claims to attempt to bring about a particular narrative outcome, but he admits that scientific papers do not simply narrate the kind of stories that he thinks are really at stake. Rather, readers assimilate papers to their own continuously revisable narrative schemes. It would be interesting to know more about how scientific articles offer themselves to this kind of reading, and how they are actually read. It would also be profitable to compare the narratives constructed by the readers of research papers with those presented in review articles and textbooks. Rouse's perspective could thereby connect with the investigations currently under way into the semiotics and hermeneutics of scientific writing, which we surveyed in Chapter 4.

For our present purposes, the important point is that historians' narratives are generally going to be at cross-purposes with scientific practitioners' own narrative understandings of the development of their fields, an observation (incidentally) made by Thomas Kuhn at the very beginning of *The Structure of Scientific Revolutions* (1962/1970). As Rouse puts it, "Scientists are situated differently from the historian because they see themselves as agents within an *unfolding* story, and consequently, that story is configured somewhat differently for them" (1990: 191). Gyorgi Markus (1987) has referred, somewhat condescendingly, to the "folk history" incorporated in scientific discourse, the function of which is dependent on its own particular kind of "amnesia." He elaborates on the workings of this folk history in pedagogy, in commemorative discourse, and in textbooks:

> Firstly, it ensures a highly individualistic picture of cognitive change in science as primarily a matter of those culture-heroes who really mattered and whose name is perpetuated. Secondly, it makes the past directly *incorporated* into the present which is seen as containing everything that is valuable (and worthy of recalling) in the past. We – pygmies or not – just stand on the shoulders of all these giants, and so we see *further* – and not *otherwise*. Thirdly, these historical *memorabilia* are also *memento mori*: science, in its relentless progress, turns even the greatest intellectual achievements into mere relicts, in it there is no other certainty besides this unlimited drive forward. (Markus 1987: 33)

Another way of construing the difference between the stories historians tell and those embedded within scientists' own discourse is suggested by a recent analysis by William Clark (1995). Clark has applied to selected writings in the history of science the categories of narrative typology which White borrowed from the literary critic Northrop Frye. Four classes of narrative are identified: Romance, Comedy, Tragedy, and Satire. Each one has a characteristic form of emplotment and typical ways of portraying the agents it describes and the scenes in which they act. Each type also communicates a particular view of the author of the work, creates a specific image of its readership, and provides a vision of the social and moral order that undergirds the story. Clark traces these six characteristic properties in four works of history of science, each of which exemplifies one of the narrative types: Charles Gillispie's *The Edge of Objectivity* (1960) ("Romance"), Shapin and Schaffer's *Leviathan and the Air-Pump* (1985) ("Tragedy"), Martin Rudwick's *The Great Devonian Controversy* (1985) ("Comedy"), and Donna Haraway's *Primate Visions* (1989) ("Satire").

Considering Gillispie's book, Clark specifies how it can be described as a romance. To characterize the book in this way, he suggests, is a more sensitive and productive reading than that which would relegate it to the disdained category of "whig history." By viewing it as a romance, we can appreciate the book's specific authorial voice, with its blend of nostalgia and apocalyptic warning. We can also see why the heroes, the great scientists who populate Gillispie's narrative, are portrayed as they are – as figures who exist largely apart from society, little sullied by contact with women or money. Heroes pursue their quests, propelled by their own genius and aided only by fortune. Their progress takes the form of blazing trails, defeating adversaries, and solving riddles, but never of mundane or routine work (Clark 1995: 9, 13, 20–23, 32–33).

Gillispie's is therefore the one of these four books that is closest to the kinds of narratives in which scientists have traditionally situated themselves, narratives that Clark (like Markus) compares with "folktales" or informal romances. Such stories are small-scale dramas "of the triumph of good over evil, of virtue over vice, of light over darkness" (White 1973: 9). They are not always simply "whiggish" – not every step taken is portrayed as a move toward the present, which Herbert Butterfield claimed was the error of the Whig constitutional historians. In fact, scientists' histories readily accommodate wrong paths mistakenly taken, obstacles overcome, and intervals when little or no progress was made. The important point is that mistakes, obstacles, and gaps can be recognized as temporary deviations and delays, particularly when the narrator himself was the hero, who now stands on the summit and traces the paths that have led to that goal.

The suggestion of the importance of romantic emplotment in scientists' portrayals of themselves and their heroes is an intriguing one. It offers fascinating prospects for historical inquiry into the construction of the figure of the scientific hero in works of biography, autobiography, and fiction. It also raises the question of the way these texts are assimilated in the processes of self-fashioning by scientists themselves. Sharon Traweek has pointed to the masculine values encoded in these "male tales," the stories of lives and careers recounted by the majority male practitioners of scientific disciplines (1988: chap. 3). They presumably fashion the narratives of their own lives to reflect stories of scientific heroes that are imparted in the course of their education and repeatedly retold in the oral culture of science. The continuing appeal of these tales may also help explain why some scientists have expressed disapproval of recent works by professional historians of science, which have frustrated their expectations that great scientists should be portrayed in a suitably heroic mode.

Somewhat disdainful of epic romance, contemporary historians of science more frequently resort to satire and irony. Whenever they surrender to their propensity to thwart "normal expectations about the kinds of resolutions provided by stories," they may be said to be indulging in satirical emplotment (White 1973: 8). And it is a normal feature of historical narrative to exhibit discontinuities of beliefs and aims that might have been disguised by actors' own reconstruals of their past. Historians relish the recovery of changes of mind and ironic reversals of fate, especially when those breaches in time have subsequently been papered over by the work of forgetting. In this respect, historians' history is always likely to differ from that of scientists. While scientists frequently emplot the history of their field so as to locate themselves and their aims in continuous relationship to it, historians are more likely to emphasize the discontinuities that divide the temporal fabric. Whereas practitioners have traditionally sought to recover a past that connects with the present, historians are more likely to seek the ironic consolation of knowledge that the past is different from the present.

This should not be taken to mean, however, that historians of science can never hope to address scientists themselves or that contemporary historical writing must inevitably lose its readers. On the contrary, there is no reason why many of them may not come to appreciate the satisfactions of alternatives to the romantic narrative mode. After all, readers of novels and viewers of films in the late twentieth century have become accustomed to a wide range of ironic, satirical, and "postmodern" narrative tricks. Readers who are sophisticated in this way would seem to be well prepared to appreciate some of the subtleties of constructivist narratives. Accounts of the making of experimental or technological systems, for example, can be plotted in terms of shifts of perspective and accommodations between actors with different outlooks and aims, as we saw in Chapters 5 and 6. Biographies can emphasize the contingencies of people's lives and the resourceful ways in which they fashion their identities to respond to them, as we saw in Chapter 2.

A certain degree of irony may even be especially appropriate to a particular narrative genre, which may expect to find readers beyond the community of professional historians: that which attempts to recapture the openness and uncertainties of scientific practice and to recount the experience of the unfolding of investigative research over the course of time. The challenge these accounts pose to the author and the reader is precisely to maintain an ironic distance from "how things turned out," so that the ways researchers revise their strategies, and even their goals, in the light of unexpected results can be perceived. The narrator has to strive to give the reader the experience of immersion in the temporal order of intellectual work, with its creative responses to continuously arising challenges and its gradual achievement of a conceptual grasp of the behavior of apparatus and material phenomena. A certain "thickness" of narrative description seems to be required to convey this unfolding experience, though the historian will also want periodically to stand back from the story, to draw attention to the discontinuities, and to point out how unexpected certain outcomes were or how they precipitated changes in the investigative approach.

Frederic L. Holmes has defended this kind of history and practised it with distinction (1987, 1992). He has drawn upon surviving laboratory notebooks of leading scientists to recount passages of their investigative practice, for instance in his study of the physiological chemistry of Antoine Lavoisier (1985). Holmes there succeeds in showing how the French chemist's famous discoveries concerning combustion and chemical composition were embedded within a research program that also concerned itself significantly with problems of respiration and fermentation, the "chemistry of life." It is to the reconstruction of this program that Holmes's book addresses itself, deploying lengthy passages of narrative with interspersed reflective comments. As Holmes insists, "we cannot shrink from telling the whole story" if we are to grasp the integrity of the investigative enterprise in which Lavoisier was engaged (1985: xx).

The results are illuminating in many respects. Readers can follow Lavoisier's frequent revisions of the ideas that guided his research, his extension of analogies from known to unknown phenomena, and his modifications of them when they failed to account for new experimental results. Even the goals that Lavoisier specified for his research were frequently altered in the light of new findings. Also well conveyed are the interactions between the different problem fields with which Lavoisier was engaged and the readiness with which he carried concepts over from one realm to another. Holmes has triumphantly vindicated his claim that close and sustained attention to Lavoisier's laboratory notebooks permits the historian to reconstruct an investigative enterprise that is only imperfectly represented in the published texts. Indeed, his scrutiny of the transition from laboratory notes to drafts to published papers has shone further light on the ways in which research findings are refined and reshaped as they are cast into a succession of written forms (Holmes 1985: 70–89, 241–251, 488).

In these respects, Holmes does ironically undercut readers' expectations of narrative continuity. His patient reconstruction of the fine-grained temporality of Lavoisier's investigative practice exhibits numerous discontinuities of interpretive frameworks and research goals. He also shows how those gaps were retrospectively smoothed over by Lavoisier's continuous redrafting of his own account of the progress of his research. The vision of ready-made science that Lavoisier offered to the world in his publications is constantly contrasted with the picture of his day-to-day practice, in which uncertainty and chance figure much more largely than they do in hindsight.

Beyond this, however, Holmes does not exploit the author's ironical advantage over his characters; he is respectful rather than satirical. Nobody could confuse a passage of his writing with one of Donna Haraway's. Lavoisier emerges from Holmes's account with his integrity intact and with the reader's respect for his creativity and his monumental labor enhanced. Holmes even, on occasion, defends Lavoisier against accusations of dishonesty, for example, in connection with his manipulations of quantitative data. He argues, quite plausibly, that Lavoisier could have invoked a variety of legitimate justifications for massaging his data as he did (1985: 190–191). The concern to protect the integrity of his subject seems to be part of what Holmes defends as the proper concern of the historian with judging the quality of scientific work (xvii). It also appears to be connected to his portrayal of Lavoisier as isolated from the social realm. Holmes marginalizes the social component of investigative practice and of Lavoisier's daily life, claiming that the world outside impinged only peripherally and occasionally upon the process of research (xvi, 501–502). The emphasis on the autonomy of the investigative process seems to go along with Holmes's insistence on his subject's integrity as a scientist.

A rather different kind of narrative account of the experience of scientific research is given by Rudwick's study of the Devonian controversy (1985). While Holmes's book is devoted to the investigative pathway pursued by an individual, Rudwick's describes the intersection of the work of a number of geologists in the course of an important debate. Thus, whereas Holmes relegates the social to the margins of his story, Rudwick builds his narrative to situate his characters in a richly described social world. It is Rudwick's declared intention, in fact, to establish the constitutive role of social interactions – along with natural phenomena – in the shaping of scientific beliefs. Toward this end, lengthy and detailed narrative is his chosen means. Because Rudwick's choice of method has attracted significant comment, it seems worth discussing at some length.

The story tells how a debate about the dating of rocks in north Devonshire in the mid-1830s led to the proposal of a new geological system, characterized by a distinct fossil fauna, which became accepted among leading British geologists in the early 1840s and was subsequently extended worldwide. Rudwick provides 340 pages of exhaustive, rigorously nonretrospective narrative which attempts to reproduce the activities and shifting claims of the main participants in the controversy on a day-by-day basis. Such a detailed account would be impossible without drawing upon an enormous wealth of archival materials, including letters, published lectures and papers, diaries, and field notebooks that have survived from this period. But, of course, the simple quantity of archival documentation is not the only justification for such lengthy narration, and Rudwick states more pressing reasons for this choice of form. "In the fine-grained study of scientific research practice, narrative is not so much a literary convenience as a methodological necessity," he proposes (Rudwick 1985: 11). Chronological narrative, at a microscopic level of resolution, offers the prospect of tracing the continuous social interactions by which scientific knowledge is shaped and hence of discriminating the constraining influences of the social and the natural worlds upon the process.

To achieve this aim, and thus to place himself on a level with the sociologists who have explored contemporary controversies in the sciences, Rudwick insists it is necessary to avoid framing the story in light of its conclusion: "Narrative in the service of understanding the shaping of knowledge must rigorously and self-consciously avoid hindsight" (1985: 12). The consistent refusal to employ the resources of retrospection is one of the most distinctive and problematic aspects of Rudwick's book. Although there are one or two instances in which actors' later reminiscences are used to fill out details of activities or to impute motives, for the most part the author refuses to ascribe significances to their findings that the participants could not themselves have recognized at the time in question. Even in regard to relatively minor issues, he declines until the very end to tell the reader how things turned out.

Such a diligently crafted narrative is no artless chronicle. The skills of the novelist were required to expand and shrink time according to the participants' experience of the flow of events. Also required was manipulation of the authorial voice, and of the role of the reader within the text, to underpin the factuality of the story and distinguish it from fiction. As regards the "inscribed author," the text presents us with a figure who is in full command of his resources and capable of firmly demarcating past events from subsequent hindsight. He even, at least half-seriously, suggests that he forgot the outcome while narrating the course of the controversy (Rudwick 1985: 12–13). This is an author in whose selectively "poor memory" readers are invited to trust.

That trust is arguably lessened once we consider the uniquely disadvantageous position in which the text situates its readers. For the crucial point is that knowledge of the future is denied not only to the actors in the narrative but also to the readers. To rule out retrospective imputations of motives and beliefs to the characters is one thing, but to deny to those following the story even the slightest knowledge of subsequent events is quite another. For Clark, this is characteristic of the teasing quality of comic narration, in which events are always unexpected. Whether because they do not like comedies, however, or because this one was simply too long, many actual readers seem to have bridled at the restriction. Rudwick even gives them a dispensation to skip portions of the narrative if they get bogged down. And, although Stephen Jay Gould suggested in his review (1987) that this ought to be illegal, he admitted that he found his own knowledge of how the dispute was ultimately resolved helped him work his way through. Many less prepared readers must have been relieved to discover that a convenient plot summary is to be found in the analytical chapters at the end of the book.

In fact, one emerges into the two concluding analytical chapters with some sense of uncertainty as to whether the long narrative journey was really necessary. Rudwick does such a brilliant job there of exhibiting the structure of the controversy, using no fewer than six different diagrammatic representations of the personal trajectories and claims of the individuals involved, that readers may well feel they understand much more from the final 60 pages than from the preceding 340. The diagrams include two comparisons of all the competing interpretations of the stratigraphical column, three (progressively more complicated) mappings of the motions of individuals through the "conceptual space" that separated their different beliefs, and one mapping of the "social space" of the geological community. The distinctive techniques of visual representation characteristic of geology itself have been put to excellent use here, as well as less obviously pertinent models such as the famous topological map of the London Underground railway system.

Certain points do unquestionably emerge more clearly from this mode of presentation than from the narrative portion of the text. The mappings make clear, for example, how large and rapid individuals' conceptual shifts sometimes were, and how frequently they coincided with experiences of fieldwork or attendance at crucial meetings. Also made evident is a point that Rudwick reiterates several times: The dispute was resolved in the social context of an intensely interactive community, but without a rhetorical or institutional victory for one side. The outcome was a consensus on a new and originally unforeseen system, not a conquest by either side, nor even a straightforward compromise. This is shown in the diagrams of conceptual space, where the Devonian system opens up as a theoretical domain formed by the crossover of the front lines of the original competing parties. Rudwick explains: "[T]he battle-lines defended by [the opposing parties], having initially faced each other in opposition, filtered silently through each other, as it were, until they faced outward, leaving at their rear a domain defended by them *both*" (1985: 405).

On the other hand, there are some important conclusions that emerge more clearly from the lengthy narrative than from the schematic analysis, though they also raise some of the problems of the way the narrative is framed. Consider the apparent difficulty of setting a date for the ending of the debate. The dispute was evidently on the wane by 1840, but, as Jack Morrell pointed out in a review (1987), it threatened to revive in the 1860s, when J. B. Jukes challenged the applicability of the Devonian system in Ireland. Instead of a rapid closure of the debate, Rudwick's narrative charts a process of gradual stabilization as the Devonian system was extended in its application across the world. By the 1860s, a local challenge was too weak to destabilize a system that had been mapped "from the Urals to Niagara."

In the relatively slow process (extending over decades) by which the Devonian system was extended and stabilized, the resolution of the controversy merged into the writing of its history. The leading participants had already shown a substantial capacity for reinterpreting their past actions and beliefs to suit their present needs. In 1839, Roderick Murchison, the original proponent of the Devonian system, conveniently forgot his previous stipulation that the Devonshire strata must contain a massive unconformity, a phenomenon that no fieldwork had found. In this sort of instance, scientists' history can be seen to be part of the work of science itself, as actors discursively maneuver to emplot the story of the past to fit their favored outcome. The historian, of course, is unlikely to go along with such a romantic mode of emplotment, and Rudwick appropriately confronts his characters with what the record shows they actually said or did.

Different problems regarding the historian's stance arise in connection with the losers of the debate, meaning here not so much Murchison's original opponent, Henry De la Beche, whose largely silent and implicit acceptance of the Devonian system Rudwick locates as precisely as he can, but the inveterate holdouts, David Williams and Thomas Weaver, who seem never to have accepted the consensus. For the sociologists Harry Collins (1987b) and Trevor Pinch (1986b), Rudwick's refusal to represent sympathetically the continued opposition to the Devonian system is a weakness that undermines his claim to have exhibited the constraining force of nature upon the participants in the controversy. If some observers refused to see things as the majority did, Collins and Pinch contend, the empirical evidence cannot have been all that compelling. For Collins, this failure to deal adequately with a few recalcitrant indi-

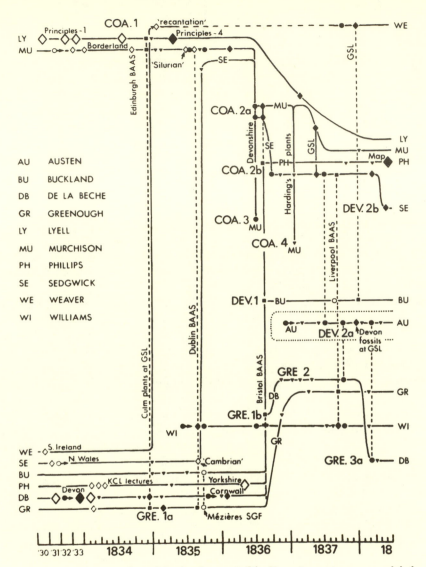

Fig. 15.5. Schematic chart of the development of the Devonian controversy, modeled on fig. 15.4 but plotting the interpretative trajectories of the ten participants most continuously involved in the controversy. The historical timescale on the horizontal axis is now quantified (but note that the years before and after the controversy are greatly compressed, to save space). The vertical axis again represents an unquantifiable impression of theoretical distance (the minor differences of level required to keep the trajectories apart have no significance). To avoid confusing an already complex chart, only the boundary of the DEV domain is shown (by a fine dotted line), using the stricter definition represented by the line of denser stippling on fig. 15.4. The historical evidence that substantiates any point on these trajectories can be found by

Figure 11. The most detailed of Martin Rudwick's mappings of the path of the Devonian controversy. Likened by reviewers to the classic map of the London Underground railway system, the chart plots the trajectories of individuals across a

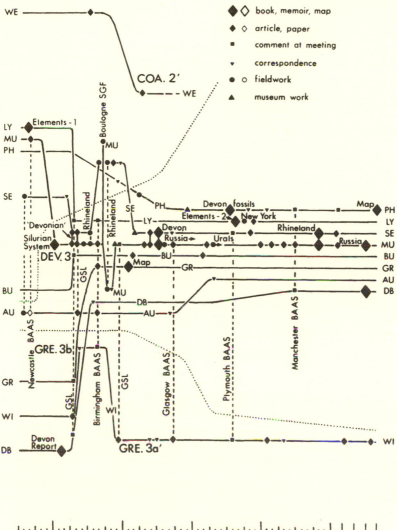

Caption to Figure 11 (*cont.*)

conceptual space in which the different theoretical positions are located. From Rudwick 1985: 412–413. Reproduced by permission of the University of Chicago Press.

viduals (the "branch lines to distant suburbs" on the "Underground map") shows that Rudwick has abandoned David Bloor's "symmetry postulate," which is taken to require equal treatment of all sides when analyzing a controversy (Collins 1987b: 826).

Rudwick's response to this criticism was that Williams and Weaver had excluded themselves from being taken seriously by their contemporaries. They were seen as having denied clear evidence in contravention of community standards (Rudwick 1988). Pinch commented that their social isolation was perhaps not irrevocable: If they had made observations in support of the mainstream opinion, they might well have been ascribed competence within the community (Pinch 1986b). But Rudwick's point seems quite valid. Pinch's counterfactual supposition cannot be put to the test, and it is hard to see what historical significance it can have.

By deploying the narrative mode, Rudwick was able to trace the temporal stabilization of facts, as they were used as resources for further investigation and their geographical and temporal reach extended. There has, in effect, been a departure from Bloorian symmetry here, but it is consequent upon the direction of the arrow of time. The historian who is following the stabilization of knowledge does not have to accept the participants' histories of the events, but, insofar as he or she is tracing the temporal resolution of disputes, the initial undertaking to treat all sides with equal sympathy will at some point have to be broken. Rudwick has provided an excellent example of the use of historical narrative to address issues of the temporality of scientific practice. The story he tells conveys a powerful impression of interpretive options being closed off, and of the rising cost of dissent from a gradually stabilizing element of knowledge. In this respect, chronological narrative does seem to make irrelevant the sociologists' objection that dissent was *in principle* still possible.

This still leaves open, however, the question of how the historian should deal with individuals who actually *did* resist the consensus. Rudwick resorts to the judgment that Williams and Weaver were no longer being taken seriously by their peers and should therefore be dropped from the story line. But this surely depends upon where the story is going. An actor who is going to succeed in reopening a controversy at a later point will need to be kept on board, unlike one who has no further role to play in the drama. Rudwick might have been more explicit than he is about how he has made these decisions. He could scarcely claim, for example, that if he had aimed to end with Jukes's challenge to the Devonian in the 1860s he would not have told the tale rather differently. This suggests the need for a somewhat more forthright specification of the author's intentions, as these are inscribed in the text. Readers are entitled to some indication of where the story is designed to end up, because this will inevitably have shaped the choice of incidents and characters and the framing of the narrative. Rudwick would have done better not to have kept his readers in the dark as to where they were being led. A somewhat more overt positioning of the author and his goals would provide concrete help in working through the narrative and would reassure readers that they were not being taken for a ride.

This is particularly important because of the debatable epistemological conclusions that Rudwick drew from his account. After the narrative and the analytical summary, he concluded with some reflections on the social construction of science and its status as natural knowledge. He tacked deftly from one side to the other of this dilemma, acknowledging, on the one hand, "the complexity, contingency, and *non*-inevitability of the scientific arguments by which [the] consensual interpretation was achieved," while, on the other, he maintained that "the Devonian can also be treated as a reliable representation of a reality that existed before it was known" (Rudwick 1985: 450, 454). Finally, he resorted to invoking the "constraints" of new empirical findings, which, while not compelling the geologists to adopt the consensus interpretation, had induced many of them to do so against their original expectations:

> In this way, it is possible to see the cumulative empirical evidence in the Devonian debate, *neither* as having determined the result of the research in any unambiguous way, as naive realists might claim, *nor* as having been virtually irrelevant to the result of the social contest on the agonistic field, as constructivists might maintain. It can be seen instead as having had a *differentiating* effect on the course and outcome of the debate, *constraining* the social construction into being a limited, but reliable and indefinitely improvable, representation of a natural reality. (Rudwick 1985: 455–456)

Andrew Pickering has raised objections to this language of constraint, and, as we saw in Chapter 1, has suggested an alternative model for understanding the unfolding of scientific practice in time (1995a, 1995b). Pickering rejects both of the alternatives between which Rudwick seems to oscillate: that knowledge is *either* constructed by purely social processes of noncompelling negotiation *or* is the revelation of a previously existing natural reality. In Pickering's scheme, scientific practice involves grappling with the material world, not just engagement with purely social entities. But it is not reduced to a process of revealing preexistent "reality." The agency of material things is captured by instrumental and human ensembles, and manifests itself as a resistance to human intentionality; it is therefore inseparable from human material practice and emerges temporally in the course of that practice. Rather than "constraint," Pickering's key metaphors are of the "mangle of practice" and the "dance of agency" by which human and material activity are mutually entwined and temporally stabilized.

It would presumably be possible to tell Rudwick's story in a way that would instantiate Pickering's account of scientific practice, though such a rewriting cannot be offered here. Rudwick's point about how the emergent outcome of the controversy was quite different from the original positions of either side reminds one of Pickering's "mangle," which reconfigures social alignments as it captures material agency. It is surely telling that Weaver and Williams found themselves excluded from the social realm of expert geologists as they persisted in their opposition to the sway of the Devonian system. It also seems significant that the temporal stabilization of the Devonian system was coincident with the geographical widening of its application. The more the work of constructing the system is grasped as a process extended in time and space, the less it seems to be simply the revelation of a preexisting reality. But, to do justice to the story in this manner would take us much too far afield. The point is that the narrative could be made to work in a rather different way, given different authorial aims. This strengthens the force of the suggestion that the author should be more willing to insert him- or herself into the story, to explain the motives that have given it its form.

To tell the story in Pickering's terms would also require some departures from strictly nonretrospective narrative. The author has to be willing to jump forward and backward in time, moving between the emergence of stable configurations of knowledge (whose stabilization is always recognized at least partially with hindsight) and the contingencies and uncertainties of the time when that stabilization could not yet be foreseen. In the narrative case studies included in his book (1995a), Pickering does this to some degree, though his stories lack Rudwick's wealth of detail and command of primary sources. His subtle model of the unfolding of scientific practice in time nonetheless challenges historians to match or refine it with their narrative powers.

The general point here is that the "naturalness" of narrative description creates both opportunities and pitfalls. Employment of the narrative mode can enable us to portray the peculiar embeddedness in time of scientific practice, but to do so requires a degree of reflexivity on behalf of the historian. Readers are entitled to know where the narrator is leading them, in order to assess how the story has been framed in light of its goal. Also, only by a departure from strict chronology can they appreciate the way in which stable knowledge emerges from uncertainty. In other words, the narrator needs to abandon the pretense of telling it just like it was and to admit to the artifactual quality of his or her narrative.

There is another reason for this, namely, to differentiate the historian's account from the self-serving narratives deployed within scientific discourse. Historians of science disavow those romantic plots that are oriented strictly toward the present, and they ought also to disavow the pretense of a purportedly unplotted narrative as a means of sustaining their own credentials (cf. Clark 1995: 57, 65). Ricoeur may be right in asserting that narrative has a certain figural dimension that connects with our experiences of temporality, but this does not mean that historians can afford to use it unreflectively. Rather, the historian's text needs to display, to some degree, "the awareness of its own inadequacy as an image of reality" (White 1973: 10). The "naturalism" of narrative can scarcely be employed unproblematically by those who claim to have shown how the "natural" is *made*.

This is particularly pertinent when we are dealing with the curious experiences of time that engagement in the making of knowledge brings about. The transition from the state of uncertainty, when it is not clear what is "real" and what is artifact, what is "signal" and what is "noise," to the subsequent crystallization of a distinction between the phenomenon and its incidental human framework is a mysterious one. In retrospect, it is most often glossed as a discontinuity in the flow of time, when the preexistent reality is suddenly revealed. Latour talks about "this miraculous emergence of new things that have always already been there" (1993: 70). To overcome this retrospective view, and to try to recapture what the experience was like, is a formidable and unsettling task, as the epigraph I have cited from the geneticist François Jacob indicates. This is nonetheless the task of a historical narrative that takes the constructivist outlook as its point of departure. Such a narrative has to register both the uncertainties of the laboring investigator and the emergence of a solution, in the light of which – looking back – everything is suddenly clear. Tracing a passage through an experienced temporality that is fractured and reflected back upon itself in this way seems to require a departure from naturalistic narrative and the model of time that it assumes. "Newtonian" time, which flows uniformly and independently of human action, does not seem adequate for such narratives (cf. Wilcox 1987). Rather, we have to chart how time itself is fragmented and realigned by the work of the construction of knowledge.

The possibilities here are intriguing, even if it is impossible even to sketch the forms they might take. Constructivist narratives that portray the distortions of temporality inherent in the project of natural science could substantially affect our understanding of the human experience of time. This would be to repay the debt to human history that constructivism incurred when it first sought to break down the barrier erected by Priestley and Whewell. Then, the strategy was to borrow from human history to illuminate the history of science. Perhaps we are now at a stage when the history of science – the history of human engagement

with the material world and of the embeddedness of knowledge in time
– can in turn shed light on the fundamental categories of human experience. Rather than regretting the passing of the comforting old stories
of scientific progress, we should therefore embrace with eagerness the
prospect of entirely new ones.

Bibliography

Abir-Am, Pnina. 1992. "A Historical Ethnography of a Scientific Anniversary in Molecular Biology: The First Protein X-Ray Photograph (1984, 1934)." *Social Epistemology* (special issue on *The Historical Ethnography of Scientific Rituals*) 6:323–354.

Abraham, Gary A. 1983. "Misunderstanding the Merton Thesis." *Isis* 74:368–387.

Achinstein, Peter, and Owen Hannaway, eds. 1985. *Observation, Experiment, and Hypothesis in Modern Physical Science*. Cambridge, MA: MIT Press.

Agar, Jon. 1994. "Regaining the Plot: Spatiality and Authority in the Historiography of Science." Unpublished typescript.

Albury, W. R. 1972. "The Logic of Condillac and the Structure of French Chemical and Biological Theory, 1780–1801." Ph.D. diss., Johns Hopkins University.

Albury, W. R., and D. R. Oldroyd. 1977. "From Renaissance Mineral Studies to Historical Geology, in the Light of Michel Foucault's *The Order of Things*." *British Journal for the History of Science* 10:187–215.

Alder, Ken. 1995. "A Revolution to Measure: The Political Economy of the Metric System in France." In Wise, ed., 39–71.

Alpers, Svetlana. 1983. *The Art of Describing: Dutch Art in the Seventeenth Century*. Chicago: University of Chicago Press.

Altick, Richard D. 1978. *The Shows of London*. Cambridge, MA: Belknap Press of Harvard University Press.

Anderson, Wilda C. 1984. *Between the Library and the Laboratory: The Language of Chemistry in Eighteenth-Century France*. Baltimore: Johns Hopkins University Press.

Appleby, Joyce, Lynn Hunt, and Margaret Jacob. 1994. *Telling the Truth about History*. New York: W. W. Norton.

Ashmore, Malcolm. 1989. *The Reflexive Thesis: Wrighting Sociology of Scientific Knowledge*. Chicago: University of Chicago Press.

Bachelard, Gaston. 1980. *La Formation de l'esprit scientifique: Contribution à une psychanalyse de la connaissance objective*. 11th ed. Paris: J. Vrin.

Bacon, Francis. 1960. *The New Organon and Related Writings*. Indianapolis: Bobbs-Merrill.

Baker, Keith Michael. 1982. "On the Problem of the Ideological Origins of the French Revolution." In *Modern Intellectual History: Reappraisals and New Perspectives*, ed. Dominick LaCapra and Steven L. Kaplan, 197–219. Ithaca, NY: Cornell University Press.

Barnes, Barry. 1974. *Scientific Knowledge and Sociological Theory.* London: Routledge and Kegan Paul.

———. 1977. *Interests and the Growth of Knowledge.* London: Routledge and Kegan Paul.

———. 1982. *T. S. Kuhn and Social Science.* London: Macmillan.

———. 1985a. *About Science.* Oxford: Basil Blackwell.

———. 1985b. "Thomas Kuhn." In *The Return of Grand Theory in the Human Sciences,* ed. Quentin Skinner, 83–100. Cambridge: Cambridge University Press.

———. 1994. "How Not to Do the Sociology of Scientific Knowledge." In Megill, ed., 21–35.

Barnes, Barry, and David Bloor. 1982. "Relativism, Rationalism and the Sociology of Knowledge." In *Rationality and Relativism,* ed. Martin Hollis and Steven Lukes, 21–47. Oxford: Basil Blackwell.

Barnes, Barry, David Bloor, and John Henry. 1996. *Scientific Knowledge: A Sociological Analysis.* Chicago: University of Chicago Press.

Barnes, Barry, and David Edge, eds. 1982. *Science in Context: Readings in the Sociology of Science.* Milton Keynes, Buckinghamshire: Open University Press.

Barnes, Barry, and Steven Shapin, eds. 1979. *Natural Order: Historical Studies of Scientific Culture.* Beverly Hills and London: Sage Publications.

Bazerman, Charles. 1987. "Literate Acts and the Emergent Social Structure of Science." *Social Epistemology* 1:295–310.

———. 1988. *Shaping Written Knowledge: The Genre and Activity of the Research Article in Science.* Madison: University of Wisconsin Press.

Bazerman, Charles, and James Paradis, eds. 1991. *Textual Dynamics of the Professions: Historical and Contemporary Studies of Writing in Professional Communities.* Madison: University of Wisconsin Press.

Bechler, Zev. 1974. "Newton's 1672 Optical Controversies: A Study in the Grammar of Scientific Dissent." In *The Interaction between Science and Philosophy,* ed. Yehuda Elkana, 115–142. Atlantic Highlands, NJ: Humanities Press.

Beer, Gillian. 1983. *Darwin's Plots: Evolutionary Narrative in Darwin, George Eliot and Nineteenth-Century Fiction.* London: Routledge and Kegan Paul.

———. 1985. "Darwin's Reading and the Fictions of Development." In Kohn, ed., 543–588.

Ben-David, Joseph. 1971/1984. *The Scientist's Role in Society: A Comparative Study.* 2d ed. Chicago: University of Chicago Press.

Benjamin, Andrew E., Geoffrey N. Cantor, and John R. R. Christie, eds. 1987. *The Figural and the Literal: Problems of Language in the History of Science and Philosophy, 1630–1800.* Manchester: Manchester University Press.

Bennett, J. A. 1986. "The Mechanics' Philosophy and the Mechanical Philosophy." *History of Science* 24:1–28.

———. 1989. "A Viol of Water or a Wedge of Glass." In Gooding, Pinch, and Schaffer, eds., 105–114.

Bennington, Geoff. 1987. "The Perfect Cheat: Locke and Empiricism's Rhetoric." In Benjamin, Cantor, and Christie, eds., 103–123.

Biagioli, Mario. 1989. "The Social Status of Italian Mathematicians, 1450–1600." *History of Science* 27:41–95.

————. 1990. "Galileo's System of Patronage." *History of Science* 28:1–62.

————. 1992. "Scientific Revolution, Social Bricolage, and Etiquette." In Porter and Teich, eds., 11–54.

————. 1993. *Galileo, Courtier: The Practice of Science in the Culture of Absolutism.* Chicago: University of Chicago Press.

Bloor, David. 1976/1991. *Knowledge and Social Imagery.* 2d ed. Chicago: University of Chicago Press.

————. 1978. "Polyhedra and the Abominations of Leviticus." *British Journal for the History of Science* 11:243–272.

————. 1983. *Wittgenstein: A Social Theory of Knowledge.* London: Macmillan.

Bono, James J. 1990. "Science, Discourse, and Literature: The Role/Rule of Metaphor in Science." In *Literature and Science: Theory and Practice,* ed. Stuart Peterfreund, 59–89. Boston: Northeastern University Press.

Borell, Merriley. 1987. "Instruments and an Independent Physiology: The Harvard Physiological Laboratory, 1871–1906." In *Physiology in the American Context, 1850–1940,* ed. Gerald L. Geison, 292–321. Bethesda, MD: American Physiological Society.

Bourdieu, Pierre. 1991. "The Peculiar History of Scientific Reason." *Sociological Forum* 6:3–26.

Bouwsma, William J. 1981. "From History of Ideas to History of Meaning." *Journal of Interdisciplinary History* 12:279–292.

Boyle, Robert. 1991. *The Early Essays and Ethics of Robert Boyle,* ed. John T. Harwood. Carbondale, IL: Southern Illinois University Press.

Brannigan, Augustine. 1981. *The Social Basis of Scientific Discovery.* Cambridge: Cambridge University Press.

Brockliss, L. W. B. 1987. *French Higher Education in the Seventeenth and Eighteenth Centuries: A Cultural History.* Oxford: Clarendon Press.

Brooke, John Hedley. 1987. "Joseph Priestley (1733–1804) and William Whewell (1794–1866): Apologists and Historians of Science. A Tale of Two Stereotypes." In *Science, Medicine and Dissent: Joseph Priestley (1733–1804),* ed. R. G. W. Anderson and Christopher Lawrence, 11–27. London: Wellcome Trust and Science Museum.

Brush, Stephen G. 1988. *The History of Modern Science: A Guide to the Second Scientific Revolution, 1800–1950.* Ames: Iowa State University Press.

Buchwald, Jed Z., ed. 1995. *Scientific Practice: Theories and Stories of Doing Physics.* Chicago: University of Chicago Press.

Bud, Robert, and Susan E. Cozzens, eds. 1992. *Invisible Connections: Instruments, Institutions, and Science.* Bellingham, WA: SPIE Optical Engineering Press.

Cahan, David. 1989. "The Geopolitics and Architectural Design of a Metrological Laboratory: The Physikalisch-Technische Reichsanstalt in Imperial Germany." In James, ed., 137–154.

Callon, Michel. 1986. "Some Elements of a Sociology of Translation: Domestication of the Scallops and the Fishermen of St. Brieuc Bay." In *Power, Action, and Belief: A New Sociology of Knowledge?* ed. John Law, 196–233. London: Routledge and Kegan Paul.

Callon, Michel, and Bruno Latour. 1992. "Don't Throw the Baby Out with the Bath School! A Reply to Collins and Yearley." In Pickering, ed., 343–368.

Cambrosio, Alberto, Daniel Jacobi, and Peter Keating. 1993. "Ehrlich's 'Beautiful Pictures' and the Controversial Beginnings of Immunological Imagery." *Isis* 84:662–699.

Cantor, Geoffrey N. 1989. "The Rhetoric of Experiment." In Gooding, Pinch, and Schaffer, eds., 159–180.

———. 1991a. "Between Rationalism and Romanticism: Whewell's Historiography of the Inductive Sciences." In *William Whewell: A Composite Portrait*, ed. Menachem Fisch and Simon Schaffer, 67–86. Oxford: Clarendon Press.

———. 1991b. *Michael Faraday: Sandemanian and Scientist. A Study of Science and Religion in the Nineteenth Century*. London: Macmillan.

Chandler, James, Arnold I. Davidson, and Harry Harootunian, eds. 1994. *Questions of Evidence: Proof, Practice, and Persuasion across the Disciplines*. Chicago: University of Chicago Press.

Christie, John R. R. 1993. "Aurora, Nemesis, and Clio." *British Journal for the History of Science* 26:391–405.

Clark, William. 1989. "On the Dialectical Origins of the Research Seminar." *History of Science* 27:111–154.

———. 1992. "On the Ironic Specimen of the Doctor of Philosophy." *Science in Context* 5:97–137.

———. 1995. "Narratology and the History of Science." *Studies in History and Philosophy of Science* 26:1–71.

Clarke, Adele E., and Joan H. Fujimura, eds. 1992. *The Right Tools for the Job: At Work in the Twentieth-Century Life Sciences*. Princeton, NJ: Princeton University Press.

Clifford, James. 1988. *The Predicament of Culture: Twentieth-Century Ethnography, Literature, and Art*. Cambridge, MA: Harvard University Press.

———. 1992. "Traveling Cultures." In *Cultural Studies*, ed. Lawrence Grossberg, Cary Nelson, and Paula A. Treichler, 96–116. New York: Routledge.

Cohen, Robert, and Thomas Schnelle, eds. 1986. *Cognition and Fact: Materials on Ludwik Fleck* (*Boston Studies in Philosophy of Science*, vol. 87). Dordrecht: Reidel.

Cole, Stephen. 1992. *Making Science: Between Nature and Society*. Cambridge, MA: Harvard University Press.

Coleman, William. 1988. "Prussian Pedagogy: Purkyne at Breslau, 1823–1839." In Coleman and Holmes, eds. 1988b, 15–64.

Coleman, William, and Frederic L. Holmes. 1988a. "Introduction." In Coleman and Holmes, eds. 1988b, 1–14.

———, eds. 1988b. *The Investigative Enterprise: Experimental Physiology in Nineteenth-Century Medicine*. Berkeley: University of California Press.

Collingwood, R. G. 1946/1961. *The Idea of History*. Oxford: Oxford University Press.

Collins, H. M. 1985. *Changing Order: Replication and Induction in Scientific Practice*. Beverly Hills and London: Sage Publications.

———. 1987a. "Certainty and the Public Understanding of Science: Science on Television." *Social Studies of Science* 17:689–713.

———. 1987b. "Pumps, Rock and Reality." *The Sociological Review* 35:819–828.

———. 1988. "Public Experiments and Displays of Virtuosity: The Core-Set Revisited." *Social Studies of Science* 18:725–748.

———. 1990. *Artificial Experts: Social Knowledge and Intelligent Machines*. Cambridge, MA: MIT Press.

Collins, Harry, and Trevor Pinch. 1993. *The Golem: What Everyone Should Know about Science*. Cambridge: Cambridge University Press.

Collins, H. M., and Steven Yearley. 1992a. "Epistemological Chicken." In Pickering, ed., 301–326.

———. 1992b. "Journey into Space." In Pickering, ed., 369–389.

Cook, Harold J. 1990. "The New Philosophy and Medicine in Seventeenth-Century England." In Lindberg and Westman, eds., 397–436.

Cooter, Roger. 1984. *The Cultural Meaning of Popular Science: Phrenology and the Organization of Consent in Nineteenth-Century Britain*. Cambridge: Cambridge University Press.

Cooter, Roger, and Stephen Pumfrey. 1994. "Separate Spheres and Public Places: Reflections on the History of Science Popularization and Science in Popular Culture." *History of Science* 32:237–267.

Darwin, Charles. 1859/1968. *The Origin of Species by Means of Natural Selection, Or the Preservation of Favoured Races in the Struggle for Life*, ed. J. W. Burrow. Harmondsworth, Middlesex: Penguin Books.

Daston, Lorraine J. 1988. "Reviews on Artifact and Experiment: The Factual Sensibility." *Isis* 79:452–470.

———. 1991. "Marvelous Facts and Miraculous Evidence in Early Modern Europe." *Critical Inquiry* 18:93–124.

Daston, Lorraine, and Peter Galison. 1992. "The Image of Objectivity." *Representations*, no. 40: 81–128.

Dear, Peter. 1985. "*Totius in Verba*: Rhetoric and Authority in the Early Royal Society." *Isis* 76:145–161.

———. 1987. "Jesuit Mathematical Science and the Reconstruction of Experience in the Early Seventeenth Century." *Studies in History and Philosophy of Science* 18:133–175.

———. 1988. "Sociology? History? Historical Sociology? A Response to Bazerman." *Social Epistemology* 2:275–278.

———. 1991a. "Narratives, Anecdotes, and Experiments: Turning Experience into Science in the Seventeenth Century." In Dear, ed., 1991b, 135–163.

———, ed. 1991b. *The Literary Structure of Scientific Argument: Historical Studies*. Philadelphia: University of Pennsylvania Press.

———. 1992. "From Truth to Disinterestedness in the Seventeenth Century." *Social Studies of Science* 22:619–631.

———. 1995a. "Cultural History of Science: An Overview with Reflections." *Science, Technology, and Human Values* 20:150–170.

———. 1995b. *Discipline and Experience: The Mathematical Way in the Scientific Revolution*. Chicago: University of Chicago Press.

Dennis, Michael Aaron. 1989. "Graphic Understanding: Instruments and Interpretation in Robert Hooke's *Micrographia*." *Science in Context* 3:309–364.

Desmond, Adrian, and James Moore. 1991. *Darwin*. London: Michael Joseph.

Dillon, George L. 1991. *Contending Rhetorics: Writing in Academic Disciplines*. Bloomington: Indiana University Press.

Durkheim, Emile. 1915/1976. *The Elementary Forms of the Religious Life*. London: George Allen and Unwin.

Eamon, William. 1984. "Arcana Disclosed: The Advent of Printing, the Books of Secrets Tradition and the Development of Experimental Science in the Sixteenth Century." *History of Science* 22:111–150.

———. 1990. "From the Secrets of Nature to Public Knowledge." In Lindberg and Westman, eds., 333–365.

———. 1991. "Court, Academy, and Printing-House: Patronage and Scientific Careers in Late Renaissance Italy." In Moran, ed. 1991b, 25–50.

———. 1994. *Science and the Secrets of Nature: Books of Secrets in Medieval and Early Modern Culture*. Princeton, NJ: Princeton University Press.

Ehrman, Esther. 1986. *Mme Du Châtelet: Scientist, Philosopher and Feminist of the Enlightenment*. Leamington Spa, Warwickshire: Berg Publishers.

Eisenstein, Elizabeth L. 1979. *The Printing Press as an Agent of Change: Communications and Cultural Transformations in Early-Modern Europe*. 2 vols. Cambridge: Cambridge University Press.

Feingold, Mordechai. 1984. *The Mathematicians' Apprenticeship: Science, Universities and Society in England, 1560–1640*. Cambridge: Cambridge University Press.

Feyerabend, Paul. 1975. *Against Method: Outline of an Anarchistic Theory of Knowledge*. London: New Left Books.

Findlen, Paula. 1989. "The Museum: Its Classical Etymology and Renaissance Genealogy." *Journal of the History of Collections* 1:59–78.

———. 1991. "The Economy of Scientific Exchange in Early Modern Italy." In Moran, ed. 1991b, 5–24.

———. 1993a. "Controlling the Experiment: Rhetoric, Court Patronage and the Experimental Method of Francesco Redi." *History of Science* 31:35–64.

———. 1993b. "Science as a Career in Enlightenment Italy: The Strategies of Laura Bassi." *Isis* 84:441–469.

———. 1994. *Possessing Nature: Museums, Collecting, and Scientific Culture in Early Modern Italy*. Berkeley: University of California Press.

Fine, Arthur. 1996. "Science Made Up: Constructivist Sociology of Scientific Knowledge." In Galison and Stump, eds., 231–254.

Fleck, Ludwik. 1935/1979. *Genesis and Development of a Scientific Fact*. Ed. Thaddeus J. Trenn and Robert K. Merton. Chicago: University of Chicago Press.

Forgan, Sophie. 1986. "Context, Image and Function: A Preliminary Enquiry into the Architecture of Scientific Societies." *British Journal for the History of Science* 19:89–113.

———. 1989. "The Architecture of Science and the Idea of a University." *Studies in History and Philosophy of Science* 20:405–434.

———. 1994. "The Architecture of Display: Museums, Universities and Objects in Nineteenth-Century Britain." *History of Science* 32:139–162.

Foucault, Michel. 1966/1970. *The Order of Things: An Archaeology of the Human Sciences*. London: Tavistock Publications.

———. 1971/1976. "The Discourse on Language." In idem, *The Archaeology of Knowledge and the Discourse on Language*, trans. A. M. Sheridan Smith, 215–237. New York: Harper and Row.

———. 1978. *Discipline and Punish: The Birth of the Prison*, trans. Alan Sheridan. New York: Random House.

———. 1983. *This Is Not a Pipe*, trans. James Harkness. Berkeley: University of California Press.

Frängsmyr, Tore, J. L. Heilbron, and Robin E. Rider, eds. 1990. *The Quantifying Spirit in the Eighteenth Century*. Berkeley: University of California Press.

Frank, Robert G., Jr. 1973. "Science, Medicine and the Universities of Early Modern England: Background and Sources." *History of Science* 11:194–216, 239–269.

Fruton, Joseph S. 1988. "The Liebig Research Group – A Reappraisal." *Proceedings of the American Philosophical Society* 132:1–66.

Fujimura, Joan H. 1992. "Crafting Science: Standardized Packages, Boundary Objects, and 'Translation.' " In Pickering, ed., 168–211.

Fuller, Steve. 1992. "Being There with Thomas Kuhn: A Parable for Postmodern Times." *History and Theory* 31:241–275.

———. 1993. *Philosophy, Rhetoric, and the End of Knowledge: The Coming of Science and Technology Studies*. Madison: University of Wisconsin Press.

———. 1995. "A Tale of Two Cultures and Other Higher Superstitions." *History of the Human Sciences* 8:115–125.

Gadamer, Hans-Georg. 1976. *Philosophical Hermeneutics*, ed. and trans. David E. Linge. Berkeley: University of California Press.

Galileo Galilei. 1954. *Dialogues Concerning Two New Sciences*, trans. Henry Crew and Alfonso de Salvio. New York: Dover Publications.

Galison, Peter. 1985. "Bubble Chambers and the Experimental Workplace." In Achinstein and Hannaway, eds., 309–373.

———. 1987. *How Experiments End*. Chicago: University of Chicago Press.

———. 1988. "History, Philosophy, and the Central Metaphor." *Science in Context* 2:197–212.

———. 1989. "In the Trading Zone." Paper presented at Tech-Know Workshop, University of California, Los Angeles, December 1989.

Galison, Peter, and Alexi Assmus. 1989. "Artificial Clouds, Real Particles." In Gooding, Pinch, and Schaffer, eds., 225–274.

Galison, Peter, and David J. Stump, eds. 1996. *The Disunity of Science: Boundaries, Contexts, and Power*. Stanford, CA: Stanford University Press.

Gascoigne, John. 1990. "A Reappraisal of the Role of the Universities in the Scientific Revolution." In Lindberg and Westman, eds., 207–260.

Geertz, Clifford. 1973. *The Interpretation of Cultures*. New York: Basic Books.

———. 1983. *Local Knowledge: Further Essays in Interpretive Anthropology*. New York: Basic Books.

Geison, Gerald L. 1978. *Michael Foster and the Cambridge School of Physiology: The Scientific Enterprise in Late Victorian Society*. Princeton, NJ: Princeton University Press.

———. 1981. "Scientific Change, Emerging Specialties, and Research Schools." *History of Science* 19:20–40.

———. 1993. "Research Schools and New Directions in the Historiography of Science." In Geison and Holmes, eds., 227–238.

———. 1995. *The Private Science of Louis Pasteur*. Princeton, NJ: Princeton University Press.

Geison, Gerald L., and Frederic L. Holmes, eds. 1993. *Research Schools: Historical Reappraisals*. *Osiris* (2d ser.) 8. Chicago: University of Chicago Press.

Gieryn, Thomas. 1988. "Distancing Science from Religion in Seventeenth-Century England." *Isis* 79:582–593.

Gillispie, Charles C. 1960. *The Edge of Objectivity: An Essay on the History of Scientific Ideas*. Princeton, NJ: Princeton University Press.

Gingras, Yves. 1995. "Following Scientists through Society? Yes, But at Arm's Length!" In Buchwald, ed., 123–148.

Goldstein, Jan. 1984. "Foucault among the Sociologists: The 'Disciplines' and the History of the Professions." *History and Theory* 23:170–192.

Golinski, Jan. 1988. "The Secret Life of an Alchemist." In *Let Newton Be!* ed. John Fauvel, Raymond Flood, Michael Shortland, and Robin Wilson, 147–167. Oxford: Oxford University Press.

———. 1989. "A Noble Spectacle: Phosphorus and the Public Cultures of Science in the Early Royal Society." *Isis* 80:11–39.

———. 1990a. "The Theory of Practice and the Practice of Theory: Sociological Approaches in the History of Science." *Isis* 81:492–505.

———. 1990b. "Language, Discourse and Science." In Olby et al., eds., 110–123.

———. 1992a. *Science as Public Culture: Chemistry and Enlightenment in Britain, 1760–1820*. Cambridge: Cambridge University Press.

———. 1992b. "The Chemical Revolution and the Politics of Language." *The Eighteenth Century: Theory and Interpretation* 33:238–251.

———. 1994. " 'The Nicety of Experiment': Precision of Measurement and Precision of Reasoning in Late Eighteenth-Century Chemistry." In *The Values of Precision*, ed. M. Norton Wise. Princeton, NJ: Princeton University Press.

Gooday, Graeme. 1990. "Precision Measurement and the Genesis of Physics Teaching Laboratories in Victorian Britain." *British Journal for the History of Science* 23:25–51.

———. 1991. " 'Nature' in the Laboratory: Domestication and Discipline with the Microscope in Victorian Life Science." *British Journal for the History of Science* 24:307–341.

Gooding, David. 1985a. " 'In Nature's School': Faraday as an Experimentalist." In *Faraday Rediscovered: Essays on the Life and Work of Michael Faraday, 1791–1867*, ed. D. Gooding and Frank A. J. L. James, 105–135. London: Macmillan.

———. 1985b. " 'He Who Proves, Discovers': John Herschel, William Pepys and the Faraday Effect." *Notes and Records of the Royal Society of London* 39:229–244.

———. 1989. " 'Magnetic Curves' and the Magnetic Field: Experimentation and Representation in the History of a Theory." In Gooding, Pinch, and Schaffer, eds., 183–223.

———. 1990. *Experiment and the Making of Meaning: Human Agency in Scientific Observation and Experiment*. Dordrecht: Kluwer Academic Publishers.

Gooding, David, Trevor Pinch, and Simon Schaffer, eds. 1989. *The Uses of Experiment: Studies in the Natural Sciences*. Cambridge: Cambridge University Press.

Gould, Stephen Jay. 1987. "The Power of Narrative." In idem, *An Urchin in the Storm: Essays about Books and Ideas*, 75–92. New York: W. W. Norton.

Greenblatt, Stephen. 1980. *Renaissance Self-Fashioning: From More to Shakespeare*. Chicago: University of Chicago Press.

Gross, Alan G. 1990. *The Rhetoric of Science*. Cambridge, MA: Harvard University Press.

Gross, Paul R., and Norman Levitt. 1994. *Higher Superstition: The Academic Left and Its Quarrels with Science*. Baltimore: Johns Hopkins University Press.

Habermas, Jürgen. 1962/1989. *The Structural Transformation of the Public Sphere: An Inquiry into a Category of Bourgeois Society*, trans. Thomas Burger with the assistance of Frederick Lawrence. Cambridge, MA: MIT Press.

Hacking, Ian. 1983. *Representing and Intervening: Introductory Topics in the Philosophy of Natural Science*. Cambridge: Cambridge University Press.

Hackmann, W. D. 1989. "Scientific Instruments: Models of Brass and Aids to Discovery." In Gooding, Pinch, and Schaffer, eds., 31–65.

Hahn, Roger. 1971. *The Anatomy of a Scientific Institution: The Paris Academy of Sciences, 1666–1803*. Berkeley: University of California Press.

Hannaway, Owen. 1975. *The Chemists and the Word: The Didactic Origins of Chemistry*. Baltimore: Johns Hopkins University Press.

———. 1986. "Laboratory Design and the Aim of Science: Andreas Libavius versus Tycho Brahe." *Isis* 77:585–610.

Haraway, Donna J. 1989. *Primate Visions: Gender, Race, and Nature in the World of Modern Science*. New York: Routledge.

———. 1991. "A Cyborg Manifesto: Science, Technology, and Socialist-Feminism in the Late Twentieth Century." In idem, *Simians, Cyborgs, and Women: The Reinvention of Nature*, 149–181. New York: Routledge.

———. 1996. "Modest Witness: Feminist Diffractions in Science Studies." In Galison and Stump, eds., 428–441.

Harris, Steven J. 1989. "Transposing the Merton Thesis: Apostolic Spirituality and the Establishment of the Jesuit Scientific Tradition." *Science in Context* 3:29–65.

Harwood, John T. 1989. "Rhetoric and Graphics in *Micrographia*." In Hunter and Schaffer, eds., 119–147.

Hodges, Andrew. 1983. *Alan Turing: The Enigma of Intelligence*. London: Unwin Paperbacks.

Holmes, Frederic L. 1985. *Lavoisier and the Chemistry of Life: An Exploration of Scientific Creativity*. Madison: University of Wisconsin Press.

———. 1987. "Scientific Writing and Scientific Discovery." *Isis* 78:220–235.

———. 1989. "The Complementarity of Teaching and Research in Liebig's Laboratory." *Osiris* (2d ser.) 5:121–164.

———. 1991. "Argument and Narrative in Scientific Writing." In Dear, ed., 164–181.

———. 1992. "Do We Understand Historically How Scientific Knowledge Is Acquired?" *History of Science* 30:119–136.

Hoskin, Keith. 1993. "Education and the Genesis of Disciplinarity: The Unexpected Reversal." In *Knowledges: Historical and Critical Perspectives on Disciplinarity*, ed. E. Messer-Davidow, D. Shumway, and D. Sylvan, 271–304. Charlottesville: University of Virginia Press.

Howell, Wilbur Samuel. 1961. *Logic and Rhetoric in England, 1500–1700*. New York: Russell and Russell.

———. 1971. *Eighteenth-Century British Logic and Rhetoric*. Princeton, NJ: Princeton University Press.

Hughes, Thomas P. 1983. *Networks of Power: Electrification in Western Society*. Baltimore: Johns Hopkins University Press.

Hunt, Bruce J. 1991. *The Maxwellians*. Ithaca, NY: Cornell University Press.

———. 1994. "The Ohm Is Where the Art Is: British Telegraph Engineers and the Development of Electrical Standards." In Van Helden and Hankins, eds., 48–63.

Hunt, Lynn, ed. 1989. *The New Cultural History*. Berkeley: University of California Press.

Hunter, Michael. 1981. *Science and Society in Restoration England*. Cambridge: Cambridge University Press.

———. 1989a. "Latitudinarianism and the 'Ideology' of the Early Royal Society: Thomas Sprat's *History of the Royal Society* (1667) Reconsidered." In Hunter 1989c, 45–71.

———. 1989b. "Promoting the New Science: Henry Oldenburg and the Early Royal Society." In Hunter 1989c, 245–260.

———. 1989c. *Establishing the New Science: The Experience of the Early Royal Society*. Woodbridge, Suffolk: Boydell Press.

Hunter, Michael, and Simon Schaffer, eds. 1989. *Robert Hooke: New Studies*. Woodbridge, Suffolk: Boydell Press.

Ihde, Don. 1991. *Instrumental Realism: The Interface between Philosophy of Science and Philosophy of Technology*. Bloomington: Indiana University Press.

Iliffe, Robert. 1989. " 'The Idols of the Temple': Isaac Newton and the Private Life of Anti-Idolatry." Ph.D. diss., Cambridge University.

———. 1992. " 'In the Warehouse': Privacy, Property and Priority in the Early Royal Society." *History of Science* 30:29–68.

Impey, Oliver, and Arthur MacGregor, eds. 1985. *The Origins of Museums: The Cabinet of Curiosities in Sixteenth- and Seventeenth-Century Europe*. Oxford: Clarendon Press.

Jacob, François. 1988. *The Statue Within: An Autobiography*, trans. Franklin Philip. New York: Basic Books.

James, Frank A. J. L., ed. 1989. *The Development of the Laboratory: Essays on the Place of Experiment in Industrial Civilization*. Basingstoke, Hampshire: Macmillan Press.

Jardine, Nicholas. 1991a. *The Scenes of Inquiry: On the Reality of Questions in the Sciences*. Oxford: Clarendon Press.

———. 1991b. "Demonstration, Dialectic, and Rhetoric in Galileo's *Dialogue*." In *The Shapes of Knowledge from the Renaissance to the Enlightenment*, ed. Donald R. Kelley and Richard H. Popkin, 101–121. Dordrecht: Kluwer Academic Publishers.

Jardine, Nicholas, J. A. Secord, and E. C. Spary, eds. 1996. *Cultures of Natural History*. Cambridge: Cambridge University Press.

Jordanova, Ludmilla. 1985. "Gender, Generation and Science: William Hunter's Obstetrical Atlas." In *William Hunter and the Eighteenth-Century Medical World*, ed. W. F. Bynum and Roy Porter, 385–412. Cambridge: Cambridge University Press.

———, ed. 1986. *Languages of Nature: Critical Essays on Science and Literature*. London: Free Association Books.

———. 1989. *Sexual Visions: Images of Gender in Science and Medicine between the Eighteenth and Twentieth Centuries*. New York: Harvester Wheatsheaf.

Keller, Evelyn Fox. 1983. *A Feeling for the Organism: The Life and Work of Barbara McClintock*. New York: Freeman.

———. 1985. *Reflections on Gender and Science*. New Haven, CT: Yale University Press.

———. 1989. "The Gender/Science System: Or, Is Sex to Gender as Nature Is to Science?" In Tuana, ed., 33–44.

———. 1992. *Secrets of Life, Secrets of Death: Essays on Language, Gender and Science*. New York: Routledge.

Kern, Stephen. 1983. *The Culture of Time and Space, 1880–1918*. Cambridge, MA: Harvard University Press.

Kim, Mi Gyung. 1992a. "The Layers of Chemical Language, I: Constitution of Bodies v. Structure of Matter." *History of Science* 30:69–96.

———. 1992b. "The Layers of Chemical Language, II: Stabilizing Atoms and Molecules in the Practice of Organic Chemistry." *History of Science* 30:397–437.

King, M. D. 1980. "Reason, Tradition, and the Progressiveness of Science." In *Paradigms and Revolutions: Applications and Appraisals of Thomas Kuhn's Philosophy of Science*, ed. Gary Gutting, 97–116. Notre Dame, IN: University of Notre Dame Press.

Knorr-Cetina, Karin. 1981. *The Manufacture of Knowledge: An Essay on the Constructivist and Contextual Nature of Science*. Oxford: Pergamon Press.

Knorr-Cetina, Karin, and Michael Mulkay, eds. 1983. *Science Observed: Perspectives on the Social Study of Science*. Beverly Hills and London: Sage Publications.

Kohler, Robert E. 1994. *Lords of the Fly: Drosophila Genetics and the Experimental Life*. Chicago: University of Chicago Press.

Kohn, David, ed. 1985. *The Darwinian Heritage*. Princeton, NJ: Princeton University Press.

Kuhn, Thomas S. 1962/1970. *The Structure of Scientific Revolutions*. 2d ed. Chicago: University of Chicago Press.

———. 1977a. *The Essential Tension: Selected Studies in Scientific Tradition and Change*. Chicago: University of Chicago Press.

———. 1977b. "Mathematical versus Experimental Traditions in the Development of Physical Science." In Kuhn 1977a, 31–65.

———. 1978. *Black-Body Theory and the Quantum Discontinuity, 1894–1912*. Oxford: Clarendon Press.

Kuklick, Henrika, and Robert Kohler, eds. 1996. *Science in the Field. Osiris* (2d ser.) 11. Chicago: University of Chicago Press.

Landes, David S. 1983. *Revolution in Time: Clocks and the Making of the Modern World*. Cambridge, MA: Harvard University Press.

Latour, Bruno. 1983. "Give Me a Laboratory and I Will Raise the World." In Knorr-Cetina and Mulkay, eds., 141–170.

———. 1986. "Visualization and Cognition: Thinking with Eyes and Hands." *Knowledge and Society: Studies in the Sociology of Culture Past and Present* 6:1–40.

———. 1987. *Science in Action: How to Follow Scientists and Engineers through Society*. Cambridge, MA: Harvard University Press.

———. 1988a. *The Pasteurization of France*. Cambridge, MA: Harvard University Press.

218   *Bibliography*

————. 1988b. "The Politics of Explanation: An Alternative." In Woolgar, ed. 1988b, 155–176.

————. 1990. "Postmodern? No, Simply Amodern! Steps Towards an Anthropology of Science." *Studies in History and Philosophy of Science* 21:145–171.

————. 1992. "One More Turn after the Social Turn . . ." In *The Social Dimensions of Science*, ed. Ernan McMullin, 272–294. Notre Dame, IN: University of Notre Dame Press.

————. 1993. *We Have Never Been Modern*, trans. Catherine Porter. Cambridge, MA: Harvard University Press.

Latour, Bruno, and Steve Woolgar. 1979/1986. *Laboratory Life: The Construction of Scientific Facts*, 2d ed. Princeton, NJ: Princeton University Press.

Leavis, F. R. 1963. *Two Cultures? The Significance of C. P. Snow*. New York: Pantheon Books.

Le Grand, Homer E. 1990a. "Is a Picture Worth a Thousand Experiments?" In Le Grand, ed. 1990b, 241–270.

————, ed. 1990b. *Experimental Inquiries: Historical, Philosophical and Social Studies of Experimentation in Science*. Dordrecht: Kluwer Academic Publishers.

Lenoir, Timothy. 1986. "Models and Instruments in the Development of Electrophysiology, 1845–1912." *Historical Studies in the Physical and Biological Sciences* 17:1–54.

————. 1988. "Practice, Reason, Context: The Dialogue between Theory and Experiment." *Science in Context* 2:3–22.

————. 1994. "Helmholtz and the Materialities of Communication." In Van Helden and Hankins, eds., 185–207.

Lenoir, Timothy, and Cheryl Lynn Ross. 1996. "The Naturalized History Museum." In Galison and Stump, eds., 370–397.

Levere, Trevor H. 1990. "Lavoisier: Language, Instruments, and the Chemical Revolution." In *Nature, Experiment, and the Sciences*, ed. Levere and W. R. Shea, 207–233. Dordrecht: Reidel.

Lewontin, Richard C. 1995. "A La Recherche du temps perdu." *Configurations* 3: 257–265.

Lindberg, David C., and Robert S. Westman, eds. 1990. *Reappraisals of the Scientific Revolution*. Cambridge: Cambridge University Press.

Livesey, Steven J. 1985. "William of Ockham, the Subalternate Sciences, and Aristotle's Theory of *Metabasis*." *British Journal for the History of Science* 59:128–145.

Locke, John. 1689/1975. *An Essay Concerning Human Understanding*, ed. Peter H. Nidditch. Oxford: Clarendon Press.

Long, Pamela O. 1991. "The Openness of Knowledge: An Ideal and Its Context in 16th-Century Writings on Mining and Metallurgy." *Technology and Culture* 32:318–335.

Lux, David S. 1991. "Societies, Circles, Academies, and Organizations: A Historiographic Essay on Seventeenth-Century Science." In *Revolution and Continuity: Essays in the History and Philosophy of Early-Modern Science*, ed. Peter Barker and Roger Ariew, 23–43. Washington, DC: Catholic University of America Press.

Lynch, Michael. 1984. *Art and Artifact in Laboratory Science: A Study of Shop Work and Shop Talk in a Research Laboratory*. London: Routledge.

————. 1985. "Discipline and the Material Form of Images: An Analysis of Scientific Visibility." *Social Studies of Science* 15:37–66.

————. 1991. "Laboratory Space and the Technological Complex: An Investigation of Topical Contextures." *Science in Context* 4:51–78.

————. 1993. *Scientific Practice and Ordinary Action: Ethnomethodology and Social Studies of Science*. Cambridge: Cambridge University Press.

Lynch, Michael, and Steve Woolgar, eds. 1990. *Representation in Scientific Practice*. Cambridge, MA: MIT Press.

MacKenzie, Donald, and Barry Barnes. 1979. "Scientific Judgment: The Biometry-Mendelism Controversy." In Barnes and Shapin, eds., 191–210.

Markus, Gyorgy. 1987. "Why Is There No Hermeneutics of Natural Sciences? Some Preliminary Theses." *Science in Context* 1:5–51.

Markus, Thomas A. 1993. *Buildings and Power: Freedom and Control in the Origin of Modern Building Types*. London: Routledge.

Marx, Karl. 1964. *The Economic and Philosophic Manuscripts of 1844*, ed. Dirk J. Struik. New York: International Publishers.

Masterman, Margaret. 1970. "The Nature of a Paradigm." In *Criticism and the Growth of Knowledge*, ed. Imre Lakatos and Alan Musgrave, 59–89. Cambridge: Cambridge University Press.

McClellan, James E. 1985. *Science Reorganized: Scientific Societies in the Eighteenth Century*. New York: Columbia University Press.

McEvoy, John G. 1979. "Electricity, Knowledge and the Nature of Progress in Priestley's Thought." *British Journal for the History of Science* 12:1–30.

Megill, Allan, ed. 1994. *Rethinking Objectivity*. Durham, NC: Duke University Press.

Mendelsohn, Everett. 1992. "The Social Locus of Scientific Instruments." In Robert Bud and Susan E. Cozzens, eds., 5–22.

Merchant, Carolyn. 1980. *The Death of Nature: Women, Ecology, and the Scientific Revolution*. San Francisco: Harper and Row.

Merton, Robert K. 1938/1970. *Science, Technology and Society in Seventeenth-Century England*, 2d ed. New York: Harper and Row.

————. 1942/1973. "The Normative Structure of Science." In Merton 1973, 267–278.

————. 1957/1973. "Priorities in Scientific Discovery." In Merton 1973, 286–324.

————. 1973. *The Sociology of Science: Theoretical and Empirical Investigations*, ed. Norman W. Storer. Chicago: University of Chicago Press.

Montgomery, Scott L. 1996. *The Scientific Voice*. New York: The Guilford Press.

Moran, Bruce T. 1991a. "Patronage and Institutions: Courts, Universities, and Academies in Germany; an Overview: 1550–1750." In Moran, ed. 1991b, 169–183.

————, ed. 1991b. *Patronage and Institutions: Science, Technology, and Medicine at the European Court 1500–1750*. Rochester, NY: Boydell Press.

Morrell, J. B. 1972. "The Chemist Breeders: The Research Schools of Liebig and Thomas Thomson." *Ambix* 19:1–46.

————. 1987. Review of Rudwick, *The Great Devonian Controversy* (1985). *British Journal for the History of Science* 20:88–89.

————. 1990. "Professionalisation." In Olby et al., eds., 980–989.

220   *Bibliography*

Mulkay, Michael. 1979. *Science and the Sociology of Knowledge.* London: Allen and Unwin.

Myers, Greg. 1990. *Writing Biology: Texts in the Social Construction of Scientific Knowledge.* Madison: University of Wisconsin Press.

———. 1992. "History and Philosophy of Science Seminar 4:00 Wednesday, Seminar Room 2: 'Fictions for Facts: The Form and Authority of the Scientific Dialogue.' " *History of Science* 30:221–247.

Nelson, John S., Allan Megill, and Donald N. McCloskey, eds. 1987. *The Rhetoric of the Human Sciences: Language and Argument in Scholarship and Public Affairs.* Madison: University of Wisconsin Press.

O'Connell, Joseph. 1993. "Metrology: The Creation of Universality by the Circulation of Particulars." *Social Studies of Science* 23:129–173.

Olby, R. C., G. N. Cantor, J. R. R. Christie, and M. J. S. Hodge, eds. 1990. *Companion to the History of Modern Science.* London: Routledge.

Oldenburg, Henry. 1965–1986. *The Correspondence of Henry Oldenburg,* ed. A. Rupert Hall and Marie Boas Hall. 13 vols. Vols. 1–9. Madison: University of Wisconsin Press. Vols. 10, 11. London: Mansell. Vols. 12, 13. London: Taylor and Francis.

Olesko, Kathryn M. 1991. *Physics as a Calling: Discipline and Practice in the Königsberg Seminar for Physics.* Ithaca, NY: Cornell University Press.

———. 1993. "Tacit Knowledge and School Formation." In Geison and Holmes, eds., 16–29.

Ong, Walter J. 1958. *Ramus, Method, and the Decay of Dialogue.* Cambridge, MA: Harvard University Press.

Ophir, Adi, and Steven Shapin. 1991. "The Place of Knowledge: A Methodological Survey." *Science in Context* 4:3–21.

Paradis, James. 1987. "Montaigne, Boyle, and the Essay of Experience." In *One Culture: Essays in Science and Literature,* ed. George Levine, 59–91. Madison: University of Wisconsin Press.

Pera, Marcello. 1988. "Breaking the Link between Methodology and Rationality: A Plea for Rhetoric in Scientific Inquiry." In *Theory and Experiment: Recent Insights and New Perspectives on Their Relation (Synthese Library,* no. 195), ed. Diderik Batens and Jean Paul van Bendegem, 259–276. Dordrecht: Reidel.

Pera, Marcello, and William R. Shea, eds. 1991. *Persuading Science: The Art of Scientific Rhetoric.* Canton, MA: Science History Publications, USA.

Pestre, Dominique. 1995. "Pour une histoire sociale et culturelle des sciences: Nouvelles définitions, nouveaux objets, nouvelles practiques." *Annales: Histoire, Sciences Sociales* 50:487–522.

Pickering, Andrew. 1984. *Constructing Quarks: A Sociological History of Particle Physics.* Edinburgh: Edinburgh University Press.

———. 1989. "Living in the Material World: On Realism and Experimental Practice." In Gooding, Pinch, and Schaffer, eds., 275–297.

———, ed. 1992. *Science as Practice and Culture.* Chicago: University of Chicago Press.

———. 1995a. *The Mangle of Practice: Time, Agency, and Science.* Chicago: University of Chicago Press.

———. 1995b. "Beyond Constraint: The Temporality of Practice and the Historicity of Knowledge." In Buchwald, ed., 42–55.

Pickstone, John V. 1993. "Ways of Knowing: Towards a Historical Sociology of Science, Technology and Medicine." *British Journal for the History of Science* 26:433–458.

———. 1994. "Museological Science? The Place of the Analytical/Comparative in Nineteenth-Century Science, Technology and Medicine." *History of Science* 32:111–138.

Pinch, Trevor J. 1986a. *Confronting Nature: The Sociology of Solar Neutrino Detection.* Dordrecht: Reidel.

———. 1986b. "Strata Various." *Social Studies of Science* 16:705–713.

———. 1992. "Opening Black Boxes: Science, Technology, and Society." *Social Studies of Science* 22:487–510.

Polanyi, Michael. 1958. *Personal Knowledge: Towards a Post-Critical Philosophy.* Chicago: University of Chicago Press.

Poovey, Mary. 1987. " 'Scenes of an Indelicate Character': The Medical 'Treatment' of Victorian Women." In *The Making of the Modern Body: Sexuality and Society in the Nineteenth Century,* ed. Catherine Gallagher and Thomas Laqueur. Berkeley: University of California Press.

Porter, Roy, and Mikuláš Teich, eds. 1992. *The Scientific Revolution in National Context.* Cambridge: Cambridge University Press.

Porter, Roy, Simon Schaffer, Jim Bennett, and Olivia Brown. 1985. *Science and Profit in 18th-Century London.* Cambridge: The Whipple Museum of the History of Science.

Porter, Theodore M. 1995. *Trust in Numbers: The Pursuit of Objectivity in Science and Public Life.* Princeton, NJ: Princeton University Press.

Prakash, Gyan. 1992. "Science 'Gone Native' in Colonial India." *Representations,* no. 40:153–178.

Prelli, Lawrence J. 1989a. *A Rhetoric of Science: Inventing Scientific Discourse.* Columbia: University of South Carolina Press.

———. 1989b. "The Rhetorical Construction of Scientific Ethos." In *Rhetoric in the Human Sciences,* ed. Herbert W. Simons, 48–68. Beverly Hills and London: Sage Publications.

Priestley, Joseph. 1767. *The History and Present State of Electricity.* London: J. Dodsley.

———. 1777. *A Course of Lectures on Oratory and General Criticism.* London: J. Johnson.

Pumfrey, Stephen. 1991. "Ideas above his Station: A Social Study of Hooke's Curatorship of Experiments." *History of Science* 29:1–44.

Ravetz, Jerome R. 1971. *Scientific Knowledge and Its Social Problems.* Oxford: Clarendon Press.

Revel, Jacques. 1991. "Knowledge of the Territory." *Science in Context* 4:133–161.

Rheinberger, Hans-Jörg. 1992a. "Experiment, Difference, and Writing, I: Tracing Protein Synthesis." *Studies in History and Philosophy of Science* 23:305–331.

———. 1992b. "Experiment, Difference, and Writing, II: The Laboratory Production of Transfer RNA." *Studies in History and Philosophy of Science* 23:389–422.

———. 1993. "Experiment and Orientation: Early Systems of In Vitro Protein Synthesis." *Journal of the History of Biology* 26:443–471.

———. 1994. "Experimental Systems: Historiality, Narration, and Deconstruction." *Science in Context* 7:65–81.

Ricoeur, Paul. 1981. *Hermeneutics and the Human Sciences: Essays on Language, Action and Interpretation,* ed. and trans. John B. Thompson. Cambridge: Cambridge University Press.

Roberts, Lissa. 1991. "Setting the Table: The Disciplinary Development of Eighteenth-Century Chemistry as Read Through the Changing Structure of Its Tables." In Dear, ed., 1991a, 99–132.

———. 1992. "Condillac, Lavoisier, and the Instrumentalization of Science." *The Eighteenth Century: Theory and Interpretation* 33:252–271.

———. 1993. "Filling the Space of Possibilities: Eighteenth-Century Chemistry's Transition from Art to Science." *Science in Context* 6:511–553.

Rorty, Richard. 1979. *Philosophy and the Mirror of Nature.* Princeton, NJ: Princeton University Press.

Rothermel, Holly. 1993. "Images of the Sun: Warren De la Rue, George Biddell Airy and Celestial Photography." *British Journal for the History of Science* 26: 137–169.

Rouse, Joseph. 1987. *Knowledge and Power: Toward a Political Philosophy of Science.* Ithaca, NY: Cornell University Press.

———. 1990. "The Narrative Reconstruction of Science." *Inquiry* 33:179–196.

———. 1991. "Philosophy of Science and the Persistent Narratives of Modernity." *Studies in History and Philosophy of Science* 22:141–162.

———. 1993a. "Foucault and the Natural Sciences." In *Foucault and the Critique of Institutions,* ed. John Caputo and Mark Yount, 137–162. University Park, PA: Pennsylvania State University Press.

———. 1993b. "What Are Cultural Studies of Scientific Knowledge?" *Configurations* 1:1–22.

———. 1996. *Engaging Science: How to Understand Its Practices Philosophically.* Ithaca, NY: Cornell University Press.

Rudwick, Martin J. S. 1976. "The Emergence of a Visual Language for Geological Science." *History of Science* 14:149–195.

———. 1985. *The Great Devonian Controversy: The Shaping of Scientific Knowledge among Gentlemanly Specialists.* Chicago: University of Chicago Press.

———. 1988. "The Closure of the Devonian Controversy." In British Society for the History of Science / History of Science Society, *Program, Papers, and Abstracts for the Joint Conference, Manchester, England, 11–15 July 1988,* 155–159.

Schaffer, Simon. 1983. "Natural Philosophy and Public Spectacle in the Eighteenth Century." *History of Science* 21:1–43.

———. 1988. "Astronomers Mark Time: Discipline and the Personal Equation." *Science in Context* 2:115–145.

———. 1989. "Glass Works: Newton's Prisms and the Uses of Experiment." In Gooding, Pinch, and Schaffer, eds., 67–104.

———. 1991. "The Eighteenth Brumaire of Bruno Latour" (review of Latour 1988a). *Studies in History and Philosophy of Science* 22:174–192.

———. 1992. "Late Victorian Metrology and Its Instrumentation: A Manufactory of Ohms." In Bud and Cozzens, eds., 23–56.

———. 1993. "The Consuming Flame: Electrical Showmen and Tory Mystics in the World of Goods." In *Consumption and the World of Goods*, ed. Roy Porter and John Brewer, 489–526. London: Routledge.

———. 1994a. "Machine Philosophy: Demonstration Devices in Georgian Mechanics." In Van Helden and Hankins, eds., 157–182.

———. 1994b. "Self Evidence." In Chandler, Davidson, and Harootunian, eds., 56–91.

Schiebinger, Londa. 1989. *The Mind Has No Sex? Women in the Origins of Modern Science*. Cambridge, MA: Harvard University Press.

———. 1993. *Nature's Body: Gender in the Making of Modern Science*. Boston: Beacon Press.

Schmitt, Charles B. 1973. "Towards a Reassessment of Renaissance Aristotelianism." *History of Science* 11:159–193.

Schuster, John, and Graeme Watchirs. 1990. "Natural Philosophy, Experiment and Discourse in the 18th Century." In Le Grand, ed. 1990b, 1–47.

Schuster, J. A., and R. R. Yeo, eds. 1986. *The Politics and Rhetoric of Scientific Method: Historical Studies*. Dordrecht: Kluwer Academic Publishers.

Secord, Anne. 1994. "Science in the Pub: Artisan Botanists in Early Nineteenth-Century Lancashire." *History of Science* 32:269–315.

Secord, James A. 1985. "Darwin and the Breeders: A Social History." In Kohn, ed., 519–542.

———. 1986. *Controversy in Victorian Geology: The Cambrian–Silurian Dispute*. Princeton, NJ: Princeton University Press.

Serres, Michel, with Bruno Latour. 1995. *Conversations on Science, Culture, and Time*. Ann Arbor: University of Michigan Press.

Shackelford, Jole. 1993. "Tycho Brahe, Laboratory Design, and the Aim of Science: Reading Plans in Context." *Isis* 84:211–230.

Shapin, Steven. 1982. "History of Science and Its Sociological Reconstructions." *History of Science* 20:157–211.

———. 1984. "Pump and Circumstances: Robert Boyle's Literary Technology." *Social Studies of Science* 14:481–520.

———. 1987. "O Henry" (review of *The Correspondence of Henry Oldenburg*, ed. A. R. Hall and M. B. Hall, 13 vols.). *Isis* 78:417–424.

———. 1988a. "Following Scientists Around" (review of Latour 1987). *Social Studies of Science* 18:533–550.

———. 1988b. "The House of Experiment in Seventeenth-Century England." *Isis* 79:373–404.

———. 1988c. "Understanding the Merton Thesis." *Isis* 79:594–605.

———. 1989. "Who Was Robert Hooke?" In Hunter and Schaffer, eds., 253–285.

———. 1990. " 'The Mind Is Its Own Place': Science and Solitude in Seventeenth-Century England." *Science in Context* 4:191–218.

———. 1991. " 'A Scholar and a Gentleman': The Problematic Identity of the Scientific Practitioner in Early Modern England." *History of Science* 29:279–327.

———. 1992. "Discipline and Bounding: The History and Sociology of Science as Seen Through the Externalism–Internalism Debate." *History of Science* 30:333–369.

———. 1993. "Personal Development and Intellectual Biography: The Case of Robert Boyle" (review of Boyle 1991). *British Journal for the History of Science* 26:335–345.

———. 1994. *A Social History of Truth: Civility and Science in Seventeenth-Century England*. Chicago: University of Chicago Press.

———. 1995. "Here and Everywhere: Sociology of Scientific Knowledge." *Annual Review of Sociology* 21:289–321.

Shapin, Steven, and Simon Schaffer. 1985. *Leviathan and the Air-Pump: Hobbes, Boyle, and the Experimental Life*. Princeton, NJ: Princeton University Press.

Shteir, Ann B. 1996. *Cultivating Women, Cultivating Science: Flora's Daughters and Botany in England 1760 to 1860*. Baltimore: Johns Hopkins University Press.

Shumway, David R., and Ellen Messer-Davidow. 1991. "Disciplinarity: An Introduction." *Poetics Today* 12:201–226.

Slaughter, Mary M. 1982. *Universal Languages and Scientific Taxonomy in the Seventeenth Century*. Cambridge: Cambridge University Press.

Smith, Adam. 1795/1980. *Essays on Philosophical Subjects*, ed. W. P. D. Wightman and J. C. Bryce. Oxford: Oxford University Press.

Smith, Crosbie, and M. Norton Wise. 1989. *Energy and Empire: A Biographical Study of Lord Kelvin*. Cambridge: Cambridge University Press.

Smith, Pamela H. 1991. "Curing the Body Politic: Chemistry and Commerce at Court, 1664–70." In Moran, ed. 1991b, 195–209.

———. 1994a. "Alchemy as a Language of Mediation at the Habsburg Court." *Isis* 85:1–25.

———. 1994b. *The Business of Alchemy: Science and Culture in the Holy Roman Empire*. Princeton, NJ: Princeton University Press.

Smith, Roger. 1992a. *Inhibition: History and Meaning in the Sciences of Mind and Brain*. Berkeley: University of California Press.

———. 1992b. "The Meaning of 'Inhibition' and the Discourse of Order." *Science in Context* 5:237–263.

Snow, C. P. 1959/1993. *The Two Cultures and the Scientific Revolution*, with intro. by Stefan Collini. Cambridge: Cambridge University Press.

Snyder, Joel. 1989. "Inventing Photography, 1839–1879." In *On the Art of Fixing a Shadow: One Hundred and Fifty Years of Photography*, ed. Sarah Greenough, Joel Snyder, David Travis, and Colin Westerbeck, 3–38. Washington, DC: National Gallery of Art.

Sprat, Thomas. 1667/1958. *The History of the Royal Society of London for the Improving of Natural Knowledge*, ed. Jackson I. Cope and Harold Whitmore Jones. Reprint, St. Louis, MO: Washington University Studies.

Stafford, Barbara Maria. 1994. *Artful Science: Enlightenment Entertainment and the Eclipse of Visual Education*. Cambridge, MA: MIT Press.

Star, Susan Leigh, and James R. Griesemer. 1989. "Institutional Ecology, 'Translations' and Boundary Objects: Amateurs and Professionals in Berkeley's Museum of Vertebrate Zoology, 1907–39." *Social Studies of Science* 19:387–420.

Stewart, Larry. 1992. *The Rise of Public Science: Rhetoric, Technology, and Natural Philosophy in Newtonian Britain, 1660–1750*. Cambridge: Cambridge University Press.

Stichweh, Rudolf. 1992. "The Sociology of Scientific Disciplines: On the Genesis and Stability of the Disciplinary Structure of Modern Science." *Science in Context* 5:3–15.

Stubbe, Henry. 1670. *Legends No Histories . . . Together with the Plus Ultra of Mr Joseph Glanvill Reduced to a Non-Plus*. London.

Stuewer, Roger H. 1985. "Artificial Disintegration and the Cambridge–Vienna Controversy." In Achinstein and Hannaway, eds., 239–307.

Terrall, Mary. 1995. "Emilie du Châtelet and the Gendering of Science." *History of Science* 33:283–310.

Thompson, E. P. 1967. "Time, Work Discipline, and Industrial Capitalism." *Past and Present*, no. 38 (December 1967): 56–97.

Tilling, Laura. 1975. "Early Experimental Graphs." *British Journal for the History of Science* 8:193–213.

Traweek, Sharon. 1988. *Beamtimes and Lifetimes: The World of High-Energy Physicists*. Cambridge, MA: Harvard University Press.

———. 1992. "Border Crossings: Narrative Strategies in Science Studies and among Physicists in Tsukuba Science City, Japan." In Pickering, ed., 429–465.

Tribby, Jay. 1991. "Cooking (with) Clio and Cleo: Eloquence and Experiment in Seventeenth-Century Florence." *Journal of the History of Ideas* 52:417–439.

———. 1992. "Body/Building: Living the Museum Life in Early Modern Europe." *Rhetorica* 10:139–163.

Tuana, Nancy, ed. 1989. *Feminism and Science*. Bloomington: Indiana University Press.

Tuchman, Arleen M. 1988. "From the Lecture to the Laboratory: The Institutionalization of Scientific Medicine at the University of Heidelberg." In Coleman and Holmes, eds. 1988b, 65–99.

Turner, Steven. 1971. "The Growth of Professorial Research in Prussia, 1818–1848 – Causes and Context." *Historical Studies in the Physical Sciences* 3:137–182.

Van Helden, Albert. 1983. "The Birth of the Modern Scientific Instrument, 1550–1700." In *The Uses of Science in the Age of Newton*, ed. John G. Burke, 49–84. Berkeley: University of California Press.

———. 1994. "Telescopes and Authority from Galileo to Cassini." In Van Helden and Hankins, eds., 9–29.

Van Helden, Albert, and Thomas L. Hankins, eds. 1994. *Instruments. Osiris* (2d ser.) 9. Chicago: University of Chicago Press.

Vickers, Brian, ed. 1987. *English Science, Bacon to Newton*. Cambridge: Cambridge University Press.

———. 1988. *In Defence of Rhetoric*. Oxford: Clarendon Press.

Warner, Deborah Jean. 1990. "What Is a Scientific Instrument, When Did It Become One, and Why?" *British Journal for the History of Science* 23:83–93.

Warwick, Andrew. 1992. "Cambridge Mathematics and Cavendish Physics: Cunningham, Campbell and Einstein's Relativity 1905–1911. Part I: The Uses of Theory." *Studies in History and Philosophy of Science* 23:625–656.

———. 1993. "Cambridge Mathematics and Cavendish Physics: Cunningham, Campbell and Einstein's Relativity 1905–1911. Part II: Comparing Traditions in Cambridge Physics." *Studies in History and Philosophy of Science* 24:1–25.

Weimar, W. 1977. "Science as a Rhetorical Transaction." *Philosophy and Rhetoric* 10:1–29.

Westman, Robert S. 1980. "The Astronomer's Role in the Sixteenth Century: A Preliminary Study." *History of Science* 18:105–147.

———. 1990. "Proof, Poetics, and Patronage: Copernicus's Preface to *De Revolutionibus*." In Lindberg and Westman, eds., 167–205.

Whewell, William. 1837/1984. *History of the Inductive Sciences*. Extracts in *Selected Writings on the History of Science*, ed. Yehuda Elkana, 1–119. Chicago: University of Chicago Press.

White, Hayden. 1973. *Metahistory: The Historical Imagination in Nineteenth-Century Europe*. Baltimore: Johns Hopkins University Press.

———. 1987. *The Content of the Form: Narrative Discourse and Historical Representation*. Baltimore: Johns Hopkins University Press.

Whitley, Richard. 1983. "From the Sociology of Scientific Communities to the Study of Scientists' Negotiations and Beyond." *Social Science Information* 22: 681–720.

Widmalm, Sven. 1990. "Accuracy, Rhetoric, and Technology: The Paris–Greenwich Triangulation, 1784–88." In Frängsmyr, Heilbron, and Rider, eds., 179–206.

Wilcox, Donald J. 1987. *The Measure of Times Past: Pre-Newtonian Chronologies and the Rhetoric of Relative Time*. Chicago: University of Chicago Press.

Williams, Mari E. W. 1989. "Astronomical Observatories as Practical Space: The Case of Pulkowa." In James, ed., 118–136.

Williams, Raymond. 1963. *Culture and Society, 1780–1950*. Harmondsworth, Middlesex: Penguin Books.

———. 1986. "Foreword." In Jordanova, ed., 10–14.

Winkler, Mary G., and Albert Van Helden. 1992. "Representing the Heavens: Galileo and Visual Astronomy." *Isis* 83:195–217.

———. 1993. "Johannes Hevelius and the Visual Language of Astronomy." In *Renaissance and Revolution: Humanists, Scholars, Craftsmen and Natural Philosophers in Early Modern Europe*, J. V. Field and Frank A. J. L. James, eds., 97–116. Cambridge: Cambridge University Press.

Wise, M. Norton (with the collaboration of Crosbie Smith). 1989a. "Work and Waste: Political Economy and Natural Philosophy in Nineteenth Century Britain (I)." *History of Science* 27:263–301.

———. 1989b. "Work and Waste: Political Economy and Natural Philosophy in Nineteenth Century Britain (II)." *History of Science* 27:391–449.

———. 1990. "Work and Waste: Political Economy and Natural Philosophy in Nineteenth Century Britain (III)." *History of Science* 28:221–261.

Wise, M. Norton, ed. 1995. *The Values of Precision*. Princeton, NJ: Princeton University Press.

Wood, Paul B. 1980. "Methodology and Apologetics: Thomas Sprat's *History of the Royal Society*." *British Journal for the History of Science* 13:1–26.

Woolgar, Steve. 1981. "Interests and Explanations in the Social Study of Science." *Social Studies of Science* 11:365–394.

———. 1988a. *Science: The Very Idea*. London: Tavistock Publishers.

———, ed. 1988b. *Knowledge and Reflexivity: New Frontiers in the Sociology of Knowledge*. Beverly Hills and London: Sage Publications.

Yeo, Richard. 1993. *Defining Science: William Whewell, Natural Knowledge, and Public Debate in Early Victorian Britain.* Cambridge: Cambridge University Press.

Young, Robert M. 1985. "Darwin's Metaphor: Does Nature Select?" In idem, *Darwin's Metaphor: Nature's Place in Victorian Culture*, 79–125. Cambridge: Cambridge University Press.

Zuckerman, Harriet. 1988. "The Sociology of Science." In *Handbook of Sociology*, ed. Niel J. Smelser, 511–574. Newbury Park, CA: Sage Publications.

Index